21世纪高等学校计算机基础实用规划教材

计算机程序设计实践教程
——C#语言

李利明　刘卫国　主编

清华大学出版社
北　京

内 容 简 介

本书是学习 C♯程序设计的教学参考书,全书包含 Visual Studio 2015 集成开发环境、实验指导、常用算法设计、习题选解和模拟试题 5 个部分。前两部分帮助读者熟悉上机环境,方便读者上机操作,通过上机实验更好地掌握 C♯程序设计的基本思想和方法。常用算法设计部分针对初学者学习程序设计的难点,总结常见问题的编程思路,帮助读者提高程序设计能力。习题选解部分通过习题练习,帮助读者复习和掌握课程内容,达到巩固和提高的目的。模拟试题部分包括 3 套模拟试题和参考答案,帮助读者检验学习效果。

本书内容丰富,实用性强,适合作为高等学校计算机程序设计课程的教学用书,也可供各类计算机应用人员阅读参考。

本书封面贴有清华大学出版社防伪标签,无标签者不得销售。
版权所有,侵权必究。侵权举报电话:010-62782989　13701121933

图书在版编目(CIP)数据

计算机程序设计实践教程:C♯语言/李利明,刘卫国主编. —北京:清华大学出版社,2018(2020.8 重印)
(21 世纪高等学校计算机基础实用规划教材)
ISBN 978-7-302-48776-0

Ⅰ. ①计… Ⅱ. ①李… ②刘… Ⅲ. ①C 语言-程序设计-高等学校-教材　Ⅳ. ①TP312.8

中国版本图书馆 CIP 数据核字(2017)第 272832 号

责任编辑:刘　星
封面设计:刘　键
责任校对:徐俊伟
责任印制:刘海龙

出版发行:清华大学出版社
　　　　　网　　址:http://www.tup.com.cn, http://www.wqbook.com
　　　　　地　　址:北京清华大学学研大厦 A 座　　　　邮　　编:100084
　　　　　社 总 机:010-62770175　　　　　　　　　　 邮　　购:010-62786544
　　　　　投稿与读者服务:010-62776969, c-service@tup.tsinghua.edu.cn
　　　　　质量反馈:010-62772015, zhiliang@tup.tsinghua.edu.cn
　　　　　课件下载:http://www.tup.com.cn,010-83470236
印 装 者:北京建宏印刷有限公司
经　　销:全国新华书店
开　　本:185mm×260mm　　　印　张:14.75　　　字　数:372 千字
版　　次:2018 年 1 月第 1 版　　　　　　　　　　　　印　次:2020 年 8 月第 2 次印刷
印　　数:1001～1200
定　　价:39.00 元

产品编号:077575-01

前 言

计算机程序设计能够体现问题求解方法,是理解计算机工作过程的有效途径,也是计算思维能力培养的重要载体。因此,计算机程序设计课程的重要性不仅体现在一般意义上的程序设计能力的培养,而且体现在引导学生实现问题求解的思维方式的转换,即学生计算思维能力的培养。在这个过程中,上机实践是十分重要的环节。

C♯语言是常用的程序实现工具,程序设计能力需要通过大量的上机实践来培养。许多程序设计方法不是光靠听课和看书就能学到的,而是通过大量的上机实践积累起来的,所以,学习程序设计不能仅限于纸上谈兵,而必须以实践为重。本书是学习C♯程序设计的教学参考书,包含5部分内容。

一是 Visual Studio 2015 集成开发环境。要上机运行一个C♯程序,需要C♯语言编译系统的支持。这部分介绍了 Visual Studio 2015 集成开发环境的使用方法,这是上机操作的基础。

二是实验指导。为方便读者上机操作,在这部分设计了12个实验,每个实验都和课程学习内容相配合,以帮助读者通过上机实验加深对课程内容的理解,更好地掌握程序设计的基本思想和方法。实验内容以编写程序练习为主,分为"模仿编程实验"和"独立编程实验"。"模仿编程实验"给出了程序的主体部分,要求将程序补充完整;"独立编程实验"则要求读者独立完成编程练习。

三是常用算法设计。面向对象程序设计的核心是从需要解决的问题中抽象出合适的类,并将数据和对数据的操作方法封装在类的内部。尽管面向对象程序设计的设计思想不同于结构化程序设计,但两者并不是对立的,在类的内部实现仍然要用到结构化程序设计的知识。所以在学习C♯程序设计时,算法设计仍然是不能忽视的问题。这部分内容根据程序设计教学基本要求,将常见的程序设计问题进行分类,分别总结每一类问题的算法设计思路,以引导读者掌握基本的程序设计方法和技巧。教学实践表明,这对提高初学者的程序设计能力是很有帮助的。

四是习题选解。这部分以课程学习为线索,编写了十分丰富的习题并给出了参考答案。在使用这些题解时,应重点理解和掌握与题目相关的知识点,而不要死记答案,应在阅读教材的基础上再来做题,通过做题达到强化、巩固和提高的目的。

五是模拟试题。这部分包括3套C♯程序设计的模拟试题和参考答案,涵盖了本课程的主要知识点,可以帮助读者了解和检验学习情况。

本书内容丰富,实用性强,适合作为高等学校计算机程序设计课程的教学用书,也可供各类计算机应用人员阅读参考。

本书第1、2、4、5章由李利明编写,第3章由刘卫国编写。此外,周肆清、周欣然、曹岳辉、蔡旭晖、李小兰、吕格莉、刘胤宏等参与了部分编写工作。清华大学出版社的工作人员对本书的策划、出版做了大量工作，在此表示衷心的感谢。

由于编者水平有限，书中难免存在不足之处，恳请广大读者批评指正。

编　者

2017 年 10 月

目 录

第 1 章 Visual Studio 2015 集成开发环境 ··· 1

1.1 Visual Studio 2015 的安装与启动 ··· 1
1.2 Visual Studio 2015 主窗口的组成 ··· 3
1.3 Visual Studio 2015 下创建 C♯控制台应用程序 ··· 5
1.4 Visual Studio 2015 程序调试 ··· 7

第 2 章 实验指导 ··· 9

实验 1 程序的运行环境和步骤 ··· 9
实验 2 C♯语言基础 ··· 11
实验 3 程序流程控制(一) ··· 14
实验 4 程序流程控制(二) ··· 18
实验 5 面向对象编程(一) ··· 22
实验 6 面向对象编程(二) ··· 29
实验 7 复杂数据表示与应用 ··· 35
实验 8 Windows 窗体与控件 ··· 40
实验 9 用户界面设计 ··· 45
实验 10 文件操作 ··· 50
实验 11 图形与图像处理 ··· 53
实验 12 数据库应用 ··· 60

第 3 章 常用算法设计 ··· 64

3.1 累加与累乘问题 ··· 64
3.2 数字问题 ··· 69
3.3 数值计算问题 ··· 73
3.4 数组的应用 ··· 76
3.5 静态方法的应用 ··· 81
3.6 解不定方程 ··· 83
思考题及答案 ··· 85

第 4 章 习题选解 ··· 88

习题 1 C♯语言概述 ··· 88

习题 2　C#程序的数据描述……………………………………………………………… 89
习题 3　程序流程控制……………………………………………………………………… 93
习题 4　面向对象编程基础……………………………………………………………… 104
习题 5　面向对象高级编程……………………………………………………………… 112
习题 6　复杂数据表示与应用…………………………………………………………… 122
习题 7　Windows 窗体与控件…………………………………………………………… 131
习题 8　用户界面设计…………………………………………………………………… 140
习题 9　文件操作………………………………………………………………………… 146
习题 10　图形与图像处理……………………………………………………………… 154
习题 11　数据库应用…………………………………………………………………… 160
参考答案…………………………………………………………………………………… 167

第 5 章　模拟试题……………………………………………………………………… 208

模拟试题 1……………………………………………………………………………… 208
模拟试题 2……………………………………………………………………………… 213
模拟试题 3……………………………………………………………………………… 218
参考答案………………………………………………………………………………… 224

参考文献…………………………………………………………………………………… 228

第 1 章　Visual Studio 2015 集成开发环境

运行 C♯ 程序需要相应编译系统的支持。Visual Studio 2015(以下简称 VS 2015)是微软公司提供的一个集成开发环境,即源程序的输入、修改及编译、运行都可以在同一环境下完成,功能齐全,操作方便。VS 2015 可以支持 C/C++、VB、Java、C♯ 编程,然而一次只能支持一种编程方式。在 VS 2015 安装完成后,第一次运行时会让用户选择常用语言,如果选择 C♯,那么就成了 C♯ 语言的编程环境,即 VC♯ 2015 集成开发环境。VS 2015 包含的功能十分丰富,本章只介绍一些常用的操作,以方便读者在 VS 2015 环境下编写 C♯ 程序。

1.1　Visual Studio 2015 的安装与启动

在启动 VS 2015 之前,首先要安装 VS 2015。VS 2015 的安装方法和其他 Windows 程序的安装方法类似。运行 VS 2015 的安装文件 setup.exe,启动安装程序后,根据屏幕提示依次确定有关内容,便可完成系统安装。

启动 VS 2015 的过程十分简单。常用的方法是,在 Windows 桌面选择"开始"|Visual Studio 2015 命令,即可启动 VS 2015 系统。启动成功后,屏幕上出现如图 1-1 所示的 VS 2015 主窗口。

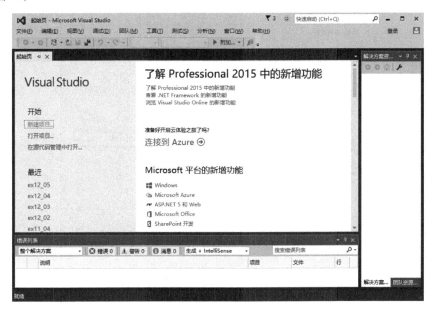

图 1-1　VS 2015 主窗口

在首次使用时,选择默认环境为"Visual C♯开发设置",这样就可在 VS 2015 中配置开发 C♯程序所需要的工具。

如果不是第一次运行 VS,在以前选择了另一个选项,为了使用 C♯环境,要把设置重置为 Visual C♯开发环境:选择"工具"|"导入导出设置"菜单,在弹出的对话框中选择"重置所有设置"选项,如图 1-2 所示。

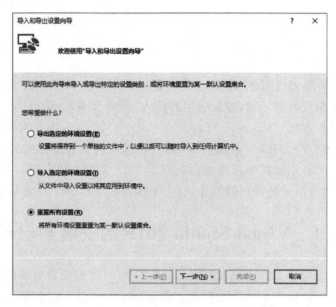

图 1-2　导入和导出设置向导——重置

单击"下一步"按钮,选择是否要保存已有的设置。如果之前对设置进行了定制,就保存设置;否则单击"否"按钮,再次单击"下一步"按钮。在下一个对话框中,选择 Visual C♯选项,如图 1-3 所示。最后单击"完成"按钮,完成设置。

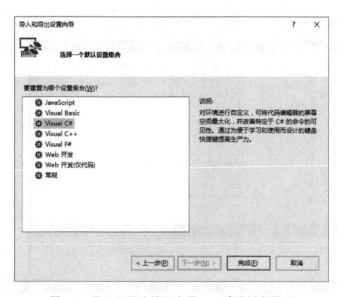

图 1-3　导入和导出设置向导——默认语言设置

VS 2015 环境布局是完全可定制的，其布局如图 1-1 所示，一般用默认设置是比较合适的。

1.2　Visual Studio 2015 主窗口的组成

和其他 Windows 程序一样，VS 2015 主窗口也具有标题栏、菜单栏和工具栏。标题栏的内容是"项目名称-Microsoft Visual Studio"。菜单栏提供了编辑、运行和调试 C♯ 程序所需要的命令。工具栏是一些命令的快捷按钮，单击工具栏上的按钮，即可执行该按钮所代表的操作。

在 VS 2015 主窗口的右侧是工作区窗口，左侧是起始页，下方是输出窗口。工作区窗口用于显示所设置的工作区的信息；起始页随着打开文件、建立项目等操作，将以分页方式显示文件，并用于输入和修改源程序；输出窗口用于显示程序编译、运行和调试过程中出现的状态信息。

1. 菜单栏

在起始状态下，VS 2015 的菜单栏共有文件(File)、编辑(Edit)、视图(View)、调试(Debug)等 10 个菜单项。在建立项目后，菜单栏共有文件(File)、编辑(Edit)、视图(View)、项目(Project)、生成(Build)、调试(Debug)等 12 个菜单项，每个菜单项都有下拉菜单，下拉菜单中的每个命令执行不同的功能。

"文件"菜单项包含用于对文件进行各种操作的命令，"编辑"菜单项包含所有与文件编辑操作有关的命令，"视图"菜单项包含用于集成开发环境各个工作窗口的显示、打开、切换的各种命令，"项目"菜单项包含用于管理项目和工作区的一系列命令，"生成"菜单项包含用于编译、连接(生成)等的命令，"调试"菜单项包含执行程序、分步执行、设立断点、建立监视项等命令。还有很多其他菜单项，实现各自功能。例如，"窗口"菜单项用于设置 VS 2015 集成开发环境中窗口的属性，"帮助"菜单项提供了详细的帮助信息。

2. 工具栏

VS 2015 集成开发环境提供了十几种工具栏。在一般情况下，系统只显示标准工具栏。要使用其他工具栏，可以通过在主窗口右击菜单栏或工具栏，在弹出的快捷菜单中选择需要显示的工具栏，或者选择"视图"|"工具栏"菜单，选择需要的工具栏。

工具栏包含很多按钮，只要把鼠标指针指向这些按钮，并且稍作停留，命令的名称就显示出来，单击这些按钮就会执行相应的操作。

3. 工作区窗口

一个 C♯ 应用程序由源程序文件、配置文件和资源文件等多个文件组成，为了更好地管理这些文件，VS 2015 引入了项目的概念。项目是由一组相互关联的文件构成的，项目的所有文件都放在项目文件夹中，项目文件夹也包含其他文件夹，这些文件夹用于保存编译和连接的输出结果。程序员通常只编写源程序，其他项目文件是由系统生成的。

VS 2015 以工作区的形式来组织文件和项目，即项目置于工作区的管理之下，因而工作区通常称为项目工作区。在 VS 2015 中工作区窗口以树状结构列出当前项目的所有文件。用户通过工作区窗口可以方便地操作这些文件。

工作区窗口通常包括 4 个选项卡，即解决方案资源管理器、类视图、属性管理器和团体资源管理器。在窗口底端单击相应图标选项卡可在 4 个选项卡之间切换，用户可以选择不同的方式操作项目。

1）解决方案资源管理器

在解决方案资源管理器中列出了项目中的所有文件和文件夹。通常一个大型的应用程序，有可能包含多种类型的项目。为了更好地组织同一个应用程序的多个项目，微软公司提出了解决方案的概念。VS 2015 的解决方案资源管理器提供了管理多个项目的能力，图1-4是包含一个项目的解决方案。当新建一个项目时，除非将项目添加到一个已存在的解决方案，否则系统将会自动生成一个新的解决方案。一个解决方案的一个项目或多个项目的所有信息保存在.sln文件中。

2）类视图

类视图用来显示当前工作区中所有类和类的成员。类视图选项卡提供了项目中所有类及其成员的层次列表。通过单击列表左侧的加号（＋）或减号（一）图标可以扩展或折叠列表。双击列表开头靠近文件夹或书本形状图标的文字也可以扩展或折叠列表。

在层次列表的每个项目前面都有一个特殊的图标。例如，保护成员或私有成员的图标是一个钥匙，全局变量是一个紫色图标。当双击某个类或成员的图标时，在编辑窗口将打开对应的代码。

用户在某一个列表项目名上右击时，将弹出一个快捷菜单，从中可以选择要执行的命令。右击的项目名不同，快捷菜单中的命令也就不同。

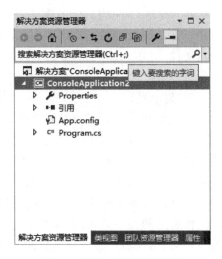

图1-4　解决方案资源管理器

3）属性管理器

在属性管理器中，可以对程序编译进行一些设置，一般采用默认设置。

4）团体资源管理器

团队资源管理器用于管理分配给用户和团队的工作，并与其他团队成员协调工作。

4. 文档窗口

在 VS 2015 中可以编辑多种不同类型的文件，如 C#源程序文件、文本文件和 HTML 文件等，每种类型的文件都具有一个默认的编辑器。当用户在解决方案资源管理器中双击相应的文件时，将使用默认的编辑器打开文件。也可在解决方案资源管理器中右击相应的文件，选择"打开方式"命令，将弹出打开方式窗口，可以在列表中选择其他的编辑器或添加新的编辑器。

VS 2015 编辑器除了具有复制、查找、替换等一般文本编辑器的功能外，还具有很多特色功能，如根据 C#语法将不同元素按照不同颜色显示，根据合适长度自动缩进等。文本编辑器还具备自动提示的功能。当用户输入程序代码时，文本编辑器会显示对应的成员函数和变量，用户可以在成员列表中选择需要的成员，这样既可以减少输入工作量，又可以避免手动输入错误。

5. 输出窗口

输出窗口主要用于显示编译、调试结果以及文件的查找信息等。

1.3　Visual Studio 2015 下创建 C♯ 控制台应用程序

1. 创建项目

在 VS 2015 主窗口中选择"文件"|"新建"|"项目"命令，这时屏幕出现"新建项目"对话框，在其左边的模板选择区域选择 Visual C♯ 选项，在中间的项目类型选择区选择"控制台应用程序"选项，如图 1-5 所示。现假设建立一个名为 ex1_01 的项目，则在"新建项目"对话框下方的名称编辑框输入 ex1_01，位置编辑框指出项目文件存放的位置，此处输入 e:\，然后单击"确定"按钮。

图 1-5　"新建项目"对话框

进入 VS 2015 主窗口，系统建立了一个解决方案的框架。解决方案资源管理器中以树状形式显示解决方案的结构，如图 1-6 所示。

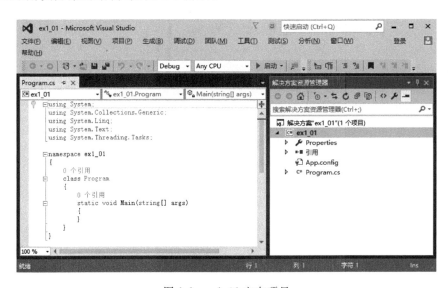

图 1-6　ex1_01 空白项目

在图 1-6 所示的解决方案管理器窗口中出现了一个名为 ex1_01 的默认解决方案,在其下方有项目 ex1_01 及其有关信息项的图标。一个解决方案可包含一个或多个项目,对于初学者来说,最好让一个解决方案只包含一个项目。VS 2015 总是将项目置于某个解决方案中进行加工处理。此时 VS 2015 在文件夹 e:\ 中生成 ex1_01 文件夹,在 e:\ex1_01 中生成多个文件,其中 ex1_01.csproj 就是项目文件,ex1_01.sln 就是解决方案文件。

2. 输入 C#源程序代码

在图 1-7 所示的界面,将光标移到左边的代码编辑窗口,在 Main 函数内输入如下程序代码(黑体字部分):

```
namespace ex1_01
{
    class Program
    {
        static void Main(string[] args)
        {
            Console.WriteLine("This is the first C# program");
        }
    }
}
```

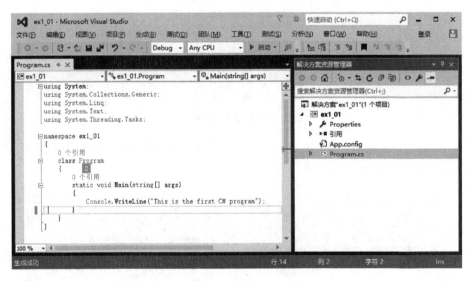

图 1-7　输入了源程序的项目窗口

在输入程序时,若有必要,可使用"编辑"菜单中的相应命令或工具栏的相应按钮。例如,要将某行注释掉或者取消掉某行的注释行时,可将光标移到相应行,再单击"注释选中行"按钮或"取消对选中行的注释"按钮进行操作。

3. 编译、连接与运行程序

在 VS 2015 主窗口中,选择"生成"|"生成解决方案"或者选择"生成"|"生成…"菜单项可将项目文件编译为可执行程序 ex1_01.exe 及执行程序所需的相关文件。生成操作也可按快捷键 F6 或者 Shift+F6 完成。文件 ex1_01.exe 及相关文件均位于文件夹 e:\ex1_01\ex1_01\bin\debug 中。

当编译或连接过程中出现错误,会在输出窗口提示错误信息,用户可根据错误提示信息对源程序进行修改,再重新生成直至没有错误信息提示。

编译、连接成功后,选择"调试"|"开始执行"命令运行 ex1_01.exe 文件,输出结果如图 1-8 所示。

执行操作也可用快捷键 F5 或 Ctrl+F5 或"标准"工具栏或"调试"工具栏中的"启动"按钮 ▶ 完成。

操作完成后,选择"文件"|"关闭解决方案"命令关闭当前解决方案,或选择"文件"|"退出"命令关闭 VS 2015。

图 1-8 ex1_01.exe 文件运行结果

与项目 ex1_01 有关的文件都保存在文件夹 e:\ex1_01 中,要复制该项目,将 e:\ex1_01 文件夹整体复制即可。

4. 打开项目

若要打开一个已存在的项目,选择"文件"|"打开"|"项目"命令,VS 2015 将弹出"打开项目"对话框。在对话框中找到项目所在的文件夹,如 e:\ex1_01,并找到解决方案文件,如 ex1_01.sln,然后单击"打开"按钮即可。

在 Windows 系统的资源管理器中找到解决方案文件,双击该文件也可打开文件所描述的项目。

1.4　Visual Studio 2015 程序调试

VS 2015 设计了一个非常方便的程序调试环境,在调试程序时可以选择调试工具对程序进行调试。下面先介绍调试工具,再介绍几种调试程序的方法。

1. "调试"菜单

当文件编辑区已经打开文件时,在"调试"菜单中选择有关命令可以以分步、跟踪方式执行程序。以下是"调试"菜单主要命令的功能。

(1) "启动调试"命令或 F5 键: 按断点(中断程序执行的位置点)分步执行程序。每执行一次"启动调试"命令或按 F5 键,程序就执行到一个断点处暂停。程序员可以检查当前变量或表达式的值。如果没有设置断点,则无停顿地执行完整个程序语句。

(2) "开始执行(不调试)"命令: 执行整个程序。

(3) "逐语句"命令: 单步执行程序,即逐条语句执行,也可按 F11 键执行该命令。当调用函数时,进入函数体内逐条语句执行。

(4) "逐过程"命令: 单步执行程序,把函数调用作为一步,即不进入函数体内执行,也可按 F10 键执行该命令。

(5) "切换断点"命令: 当光标移到要设置断点的代码行时,选择该命令或按 F9 键,设置一个断点。这时,可以看到代码窗口左边的断点区出现一个红点。重复此操作则取消断点。设置断点更简单的方法是直接单击代码窗口的断点区。

(6) "新建断点"命令: 在函数内部设置断点。

2. "调试"工具栏

"调试"工具栏提供"调试"菜单主要命令的快捷方式,共包括12个按钮,如图1-9所示。用户可以使用该工具栏上的按钮及其快捷键调试程序。

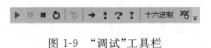

图1-9 "调试"工具栏

3. 程序调试举例

图1-10 显示了一个程序的调试过程。

图1-10 程序的调试

1) 设置断点

在程序中某个关键的语句处设置断点,最简捷的方法是,单击要设置断点的语句行左边的断点区,这时,行首断点区出现一个红色的圆点。再次单击将取消断点。

2) 开始调试

选择"启动调试"命令或按F5键,程序进入调试状态。若进入调试状态之前程序没有经过编译,系统首先会进行编译。没有编译错误的程序才可进行调试。

进入调试后,程序执行到第一个断点处暂停,这时可以查看有关的中间结果。在图1-10中,左下角窗口中系统自动显示了有关变量的值,其中,n、s、t的值分别是3、4、6。图中左侧断点区的箭头表示当前程序暂停的位置。

再继续按F5键,则执行到下一断点处暂停,若无下一断点,则将执行完整个程序。若再继续按F11键则接着进行"逐语句"调试。

3) 停止调试

在调试过程中,可选择"调试"|"停止调试"命令或按Shift+F5键结束调试,回到初始状态。

第 2 章　实验指导

学习程序设计,上机实验是十分重要的环节。为了方便读者上机练习,本章设计了 12 个实验。这些实验和课堂教学紧密配合,读者可以根据实际情况从每个实验中选择部分内容作为上机练习,各个实验后面的实验思考题可以作为实验内容的补充。

为了达到理想的实验效果,读者务必做到:

(1) 实验前认真准备,要根据实验目的和实验内容,复习好实验中要用到的概念、语句,想好编程的思路,做到胸有成竹,提高上机效率。

(2) 实验过程中积极思考,要深入分析实验现象、程序的执行结果以及各种屏幕信息的含义、出现的原因并提出解决办法。

(3) 实验后认真总结,要总结本次实验有哪些收获,还存在哪些问题,并写出实验报告。实验报告应包括实验目的、实验内容、流程图、程序清单、运行结果以及实验的收获与体会等内容。

实验 1　程序的运行环境和步骤

一、实验目的

1. 熟悉 VS 2015 集成开发环境的使用方法。
2. 掌握运行一个 C♯ 程序的基本步骤。
3. 熟悉 C♯ 程序的结构特征和程序书写规则。

二、模仿编程实验

1. 在 VS 2015 环境下,练习程序的编辑、生成和运行。

```
using System;
namespace ex01_01
{
    class Program
    {
        static void Main(string[] args)
        {
            Console.WriteLine("Hello, C♯!");
        }
    }
}
```

2. 下面是一个减法程序,程序运行时等待用户从键盘输入两个整数,然后求它们的差,上机运行该程序。

```csharp
using System;
namespace ex01_02
{
    class Program
    {
        static void Main(string[] args)
        {
            int a, b, c;
            Console.Write("a:");
            a = int.Parse(Console.ReadLine());
            Console.Write("b:");
            b = int.Parse(Console.ReadLine());
            c = a - b;
            Console.WriteLine("a-b={0}", c);
        }
    }
}
```

3. 计算并输出半径为 5 的圆的面积。请补充程序,并上机运行该程序。

```csharp
using System;
namespace ex01_03
{
    class Program
    {
        static void Main(string[] args)
        {
            double p;
            p = ____①____ ;
            Console.WriteLine("半径为 5 的圆的面积:{0}", p);
        }
    }
}
```

4. 分析下面的程序,改正其中的错误。

```csharp
using System;
namespace ex01_04
{
    class Program
    {
        static void Main(string[] args)
        {
            int i = 15, j = 6;
            s = i + j;                            /* 变量 s 没有定义 */
            Console.WriteLine("s={0}", s);
        }
    }
}
```

回答下列问题：

（1）生成时出现的编译信息是什么？分析编译信息的含义，修改错误后再生成程序。

（2）生成并运行程序，分析输出结果，说出产生这种结果的原因是什么？

三、独立编程实验

1. 先输入并运行一个最简单的 C♯ 程序，然后对该程序进行扩充，使其能输出简单的一串字符或数值，构成比较简单的 C♯ 程序。再进一步思考，这一类程序在调试程序、验证系统时有何价值？

提示：最简单的 C♯ 程序是指程序中的内容是语法上必须要求的，哪怕缺少任何一个字符都会出现语法错误。

2. 从键盘输入任意 3 个整数，求它们的和及平均值。

3. 输出下列图案。

```
    *
   ***
  *****
 *******
```

实验 2 C♯ 语言基础

一、实验目的

1. 熟悉 C♯ 基本数据类型和常量的表示方法、变量的定义与使用规则。
2. 掌握 C♯ 的有关运算符的运算规则与表达式的书写方法。
3. 掌握简单的数据输出、输入方法。
4. 熟悉不同类型数据运算时数据类型的转换规则。

二、模仿编程实验

1. 阅读以下程序，写出它们的输出结果并上机验证。

（1）

```csharp
using System;
namespace ex02_01
{
    class Program
    {
        static void Main(string[] args)
        {
            int m = 15, n = 2, y;
            double x;
            x = m / n + (double)m / n + 1 / n;      //不同类型数据运算
            y = (int)x;
            Console.WriteLine("{0}\t{1}", x, y);
        }
```

        }
    }

(2)
```
using System;
namespace ex02_02
{
    class Program
    {
        static void Main(string[ ] args)
        {
            char c, h;
            int i, j;
            c = 'a';
            h = (char)( c - ('z' - 'Z'));           //小写变大写
            i = 'a';
            j = 'a' + 1;
            Console.WriteLine("{0},{1},{2},{3},{4}", c, h, (int)c, i, j);
        }
    }
}
```

(3)
```
using System;
namespace ex02_03
{
    class Program
    {
        static void Main(string[ ] args)
        {
            int i = 5, j = 7, k, l, m = 15, n = -26;
            k = ++i;
            l = j++;
            m += i++;
            n -= --j;
            Console.WriteLine("{0}{1}{2}{3}{4}{5}",i,j,k,l,m,n);
        }
    }
}
```

思考：如果要将输出的数据用逗号分隔开，应如何修改程序？

(4)
```
using System;
namespace ex02_04
{
    class Program
    {
        static void Main(string[ ] args)
        {
```

```
            Console.WriteLine("{0} {1}", 'a', 'b');
            Console.WriteLine("{0}\t{1}\n", 'd','e');
            Console.WriteLine(" * \n ** \n *** \n **** ");
        }
    }
}
```

2. 阅读程序,改正其中的语法错误,然后运行程序,观察输出内容的格式,说明各输出语句的功能。

```
using System;
namespace ex02_05
{
    class Program
    {
        static void Main(string[ ] args)
        {
            string myFName = 'Tom';
            string myLName = 'Cruise';
            int myInt = 100
            Console.WriteLine("First Name = |{0,10}|", myFName);
            Console.WriteLine("Last Name = |{0,10}|", myLName);
            Console.WriteLine("Price = |{0,10:C}|", myInt);
            Console.WriteLine("First Name = |{0, -10}|", myFName);
            Console.WriteLine("Last Name = |{0, -10}|", myLName);
            Console.WriteLine("Price = |{0, -10:C}|", myInt);
        }
    }
}
```

3. 输入两个整数,分别存入 a 和 b 变量,不借助于第三个变量,交换 a 和 b 变量的值并输出。请完善程序。

```
using System;
namespace ex02_06
{
    class Program
    {
        static void Main(string[ ] args)
        {
            int a, b;
            _____①_____ ;
            b = int.Parse(Console.ReadLine());
            Console.WriteLine("交换前: a = {0}; b = {1}", a,b);
            a = a + b;
            _____②_____ = a - b;
            _____③_____ = a - b;
            Console.WriteLine("交换后: a = {0}; b = {1}", _____④_____ );
        }
    }
}
```

4. 阅读以下程序,写出它们的输出结果并上机验证。

```csharp
using System;
namespace ex2_07
{
    class Program
    {
        static void Main(string[] args)
        {
            int i1 = 68, i2, i3;
            double d1 = 69.125,d2,d3;
            char c1 = 'C', c2, c3;
            Console.WriteLine("i1 = {0:d5},d1 = {1:f},c1 = {2}", i1, d1, c1);
            i2 = (int)d1;
            d2 = i1;
            c2 = (char)i1;
            Console.WriteLine("i2 = {0:d5},c2 = {1:f},d2 * c2 = {2}", i2,d2, d2 * c2);
            i3 = c1;
            d3 = (int)d1;
            c3 = (char)d1;
            Console.WriteLine("i3 = {0:d5},i3 * d3 = {1:f},c3 % 2 = {2}", i3, i3 * d3, c3 % 2);
        }
    }
}
```

三、独立编程实验

1. 已知 a=2,b=7,x=7.9,y=3.2(a,b 为整型,x,y 为浮点型),求算术表达式(float)(a+b)/2+(int)x%(int)y+b/a 的值,并上机验证。
2. 输入字母字符,输出其对应的 ASCII 码值。
3. 输入某大写字母字符,输出其对应的小写字母。
4. 从键盘输入一个三位整数,分别输出其个位、十位、百位数字。
5. 输入直角三角形的一条直角边长和一条斜边,求另一直角边的长度。

实验 3　程序流程控制(一)

一、实验目的

1. 掌握关系运算、逻辑运算。
2. 熟悉 if 语句的各种形式并掌握用 if 语句实现选择结构的方法。
3. 熟悉 switch 语句的格式并掌握实现多分支选择结构的方法。

二、模仿编程实验

1. 阅读以下程序,写出它们的输出结果并上机验证。

（1）
```
using System;
namespace ex03_01
{
    class Program
    {
        static void Main(string[] args)
        {
            int a, b, c;
            bool x;
            a = 15;
            b = 18;
            c = 21;
            x = a < b || c++ > 0;
            Console.WriteLine("x = {0},c = {1}", x, c);
        }
    }
}
```

（2）
```
using System;
namespace ex03_02
{
    class Program
    {
        static void Main(string[] args)
        {
            int a = 3, b = 2, c = 3;
            bool d;
            Console.WriteLine("c = [a<b?a:b] = {0}", c += a > b ? ++a : ++b);
            Console.WriteLine("a = {0},b = {1}", a, b);
            Console.WriteLine("d = {0}", d = --a < b++);
            Console.WriteLine("a = {0},b = {1}", a, b);
            Console.WriteLine("d = {0}", d = ++a > b-- || ++b > a);
            Console.WriteLine("a = {0},b = {1}", a, b);
        }
    }
}
```

2. 下面程序功能是从键盘输入一个实数,然后输出其绝对值。请完善程序。

```
using System;
namespace ex03_03
{
    class Program
    {
        static void Main(string[] args)
        {
            double x;
            Console.WriteLine("请输入一个实数: ");
```

```
            x = double.Parse(_____①_____);
            if (x < 0)
                _____②_____;
            Console.WriteLine("输入数据的绝对值为:{0}", x);
        }
    }
}
```

3. 下面程序判断输入的字符是否为英文字母,若是,则输出 yes 信息和该字母的 ASCII 码;否则提示 no。请完善程序,并回答问题。

```
using System;
namespace ex03_04
{
    class Program
    {
        static void Main(string[] args)
        {
            char c;
            Console.WriteLine("请输入一个字符: ");
            c = char.Parse(_____①_____);
            if ((c >= 'A' && c <= 'Z') || (_____②_____))
            {
                Console.WriteLine("yes\n");
                Console.WriteLine(" ASCII 码值:{0}", _____③_____);
            }
            else
                _____④_____;
        }
    }
}
```

思考:如果删除题中 if 语句里的一对花括号,结果会怎样?

4. 下列程序的功能是输入三个整数,求三个数中的最小值。请完善程序。

```
using System;
namespace ex03_05
{
    class Program
    {
        static void Main(string[] args)
        {
            int num1, num2, num3, min;
            Console.WriteLine("请输入第一个数: ");
            num1 = int.Parse(Console.ReadLine());
            Console.WriteLine("请输入第二个数: ");
            num2 = int.Parse(Console.ReadLine());
            Console.WriteLine("请输入第三个数: ");
            num3 = int.Parse(Console.ReadLine());
            if (num1 < num2)
                min = _____①_____;
```

```
        else
            min = ____②____ ;
        if ( ____③____ )
            min = num3;
        Console.WriteLine("三个整数中的最小值 min = {0}", min);
        }
    }
}
```

5. 下列多分支选择结构程序的功能是根据输入的年、月,判断该月的天数。根据历法,闰年 2 月的天数为 29 天,其他年份 2 月为 28 天。闰年是指年份可以被 4 整除而不能被 100 整除,或者能被 400 整除的年份。请完善程序。

```
using System;
namespace ex03_06
{
    class Program
    {
        static void Main(string[ ] args)
        {
            ushort year, month, days = 0;
            Console.WriteLine("请输入年:");
            year = ushort.Parse(Console.ReadLine());
            Console.WriteLine("请输入月:");
            month = ushort.Parse(Console.ReadLine());
            switch (month)
            {
                case 1:
                case 3:
                case 5:
                case 7:
                case 8:
                case 10:
                case 12: days = 31; ____①____ ;
                case 4:
                case 6:
                case 9:
                case 11: days = 30; ____②____ ;
                case 2:
                    if ( ____③____ )
                        days = 29;
                    else
                        days = 28;
                    break;
            }
            Console.WriteLine("{0}年{1}月的天数为:{2}", year, month, days);
        }
    }
}
```

三、独立编程实验

1. 输入三个正实数,判断能否以它们为边长构成三角形,若能,求其面积。

$$s=\sqrt{p(p-a)(p-b)(p-c)},\ p=\frac{a+b+c}{2}$$,其中 a、b、c 表示边长。

2. 输入两个字符,若这两个字符的 ASCII 码之差为偶数,则输出它们的后一个字符,否则输出它们的前一个字符。

3. 输入 x 的值,求相应的 y=f(x)值。分段函数 f(x):

$$y = f(x) = \begin{cases} \cos x + \sqrt{x^2+1} & (x \neq 0) \\ \sin x - x^2 + 3x & (x = 0) \end{cases}$$

4. 给出一个百分制成绩,要求输出成绩等级 A,B,C,D,E。90 分以上为 A,80~89 分为 B,70~79 分为 C,60~69 分为 D,60 分以下为 E。当输入数据大于 100 或小于 0 时,通知用户"输入数据出错",程序结束。要求分别用 if 语句和 switch 语句实现。

实验 4 程序流程控制(二)

一、实验目的

1. 掌握 while、do-while 和 for 循环语句的格式和使用方法。
2. 掌握循环结构程序设计的方法。
3. 理解 break 和 continue 语句在循环结构中的不同作用。

二、模仿编程实验

1. 阅读以下程序,写出输出结果并上机验证。

(1)
```
using System;
namespace ex04_01
{
    class Program
    {
        static void Main(string[] args)
        {
            int a = 3, b = 7;
            for (; a <= 10; a++)
            {
                if (b >= 9) break;
                if (b % 2 == 0) { b += 5; continue; }
                b--;
            }
            Console.WriteLine("a={0},b={1}", a, b);
        }
    }
}
```

（2）运行程序，写出分别输入 123、13a、234 及 123 的结果。

```csharp
using System;
namespace ex04_02
{
    class Program
    {
        static void Main(string[] args)
        {
            try
            {
                int sum, num1, num2;
                string s1, s2;
                s1 = Console.ReadLine();
                s2 = Console.ReadLine();
                num1 = convert(s1);
                num2 = convert(s2);
                sum = num1 + num2;
                Console.WriteLine("{0} + {1} = {2}", num1, num2, sum);
            }
            catch (FormatException ex)
            {
                Console.WriteLine("您的输入不符合要求,数据转换异常");
            }
        }
        static int convert(string s)
        {
            try
            {
                return Convert.ToInt32(s);
            }
            catch (Exception ex)
            {
                Console.WriteLine("抛出异常");
                throw;             //如果将 throw 改为 return 0,会是什么结果?
            }
        }
    }
}
```

2. 从键盘输入数据，若输入的数据是负数则程序运行结束。求数据中的最大数、最小数和平均数。请完善程序。

```csharp
using System;
namespace ex04_03
{
    class Program
    {
        static void Main(string[] args)
        {
            float m, max, min, s = 0;
```

```
        int n = 0;
        Console.WriteLine("输入数 m:");
        m = float.Parse(Console.ReadLine());
        max = min = ____①____ ;
        s = m; n = 1;
        while ((m = float.Parse(Console.ReadLine())) >= 0)
        {
            if (m > max)
                ____②____ ;
            if (m < min)
                ____③____ ;
            s = ____④____ ;
            n = n + 1;
        }
        Console.WriteLine("最大数 = {0}", max);
        Console.WriteLine("最小数 = {0}", min);
        Console.WriteLine("平均 = {0}", s/n);
    }
  }
}
```

思考：若将题中的 while 语句改成 while(m>=0)，该怎么修改程序？

3. 计算π的近似值，直到最后一项的绝对值小于 10^{-8} 为止。π的近似计算公式为

$$\frac{\pi}{4} \approx 1 - \frac{1}{3} + \frac{1}{5} - \frac{1}{7} + \cdots$$

请改正程序中的错误。

```
using System;
namespace ex04_04
{
    class Program
    {
        static void Main(string[] args)
        {
            double f = 1, sum = f;
            int d = 1;
            int sign = 1;
            do
            {
                d += 2;
                sign = -sign;
                f = 1 /d;                       //此行有错
                sum += sign * f;
            }while (f < 1e-8);                  //此行有错
            sum *= 4;
            Console.WriteLine("PI≈{0}", sum);
        }
    }
}
```

4. 已知 ASCII 码值在 32～126 中的字符可以显示输出，其余码值的字符为不可显示的

控制字符。下面程序将可显示的字符制成表格输出,并使每个字符与它的 ASCII 码值对应起来,每行输出 7 个字符。请完善程序。

```
using System;
namespace ex04_05
{
    class Program
    {
        static void Main(string[] args)
        {
            ushort i = 0, asci;                    //i控制每行的字符数
            char c;
            Console.WriteLine("\t          ASCII 码对照表\n");
            for (asci = 32; asci <= 126;  ①    )
            {
                c = (char)asci;
                Console.Write("{0} = {1}\t", c, asci);
                    ②    ;
                if (i == 7)
                {
                    i = 0;
                        ③    ;
                }
            }
            Console.Write("\n");
        }
    }
}
```

5. 求 1000 内所有的完数。所谓"完数"是指与其因子之和相等的数(除本身之外)。例如,6=1+2+3,而 1、2 和 3 都是 6 的因子。要求以如下形式输出:6-->1,2,3。

```
using System;
namespace ex04_06
{
    class Program
    {
        static void Main(string[] args)
        {
            int i, j, sum;
            for (i = 2; i <= 1000; i++)
            {     ①    ;
                for (j = 1; j <= i / 2; j++)        //求 i 的因子和
                    if (i % j == 0) sum = sum + j;
                if (    ②    )                      //判断 i 是否为完数
                {
                    Console.Write("{0} --->1", i);
                    for (j = 2; j <= i / 2; j++)    //按指定格式输出完数
                        if (    ③    )
                            Console.Write(",{0}", j);
                    Console.Write("\n");
```

                }
            }
        }
    }
}

三、独立编程实验

1. 输出 1900—2050 年内所有闰年的年份。

2. 计算 s=1+(1+2)+(1+2+3)+(1+2+3+4)+…+(1+2+3+…+n)的值，n 值从键盘输入。

3. 求数列 1/2,2/3,3/5,5/8,8/13,12/21,…前 n 项之和。

4. 求[1,100]间有偶数个不同因子的整数个数及其中最大的一个数。

5. 设 abcd×e=dcba(a 非 0,e 非 0 非 1),求满足条件的 abcd 与 e。

实验 5　面向对象编程（一）

一、实验目的

1. 掌握类和对象的定义和使用。
2. 掌握构造函数和析构函数的作用、定义方式和实现方法。
3. 掌握类中静态方法的定义和设计。
4. 掌握类中属性和索引器的设计。

二、模仿编程实验

1. 阅读下列程序，在注释中写出输出结果并上机验证。

(1)

```
using System;
namespace ex05_01
{
    public class TData
    {
        int x, y;
        public void setdata(int x1,int y1)
        {
            x = x1;y = y1;
        }
        public void dispdata(string s)
        {
            Console.WriteLine("{0}:{1}--{2}",s, x, y);
        }
    }
    class Program
    {
        static void Main(string[] args)
```

```
        {
            TData d1 = new TData(), d2;
            d1.setdata(5, 1);
            d1.dispdata("d1");                  //输出_____
            d2 = new TData();
            d2.setdata(7, 9);
            d2.dispdata("d2");                  //输出_____
        }
    }
}
```

(2)
```
using system;
namespace ex05_02
{
    public class TData
    {
        int x = 0, y = 0;
        public TData() { }
        public TData(int x1, int y1)
        {
            x = x1; y = y1;
        }
        public void dispdata(string s)
        {
            Console.WriteLine("{0}:{1} -- {2}", s, x, y);
        }
    }
    class Program
    {
        static void Main(string[] args)
        {
            TData d1 = new TData(2,3), d2;
            d1.dispdata("d1");                  //输出_____
            d2 = new TData();
            d2.dispdata("d2");                  //输出_____
        }
    }
}
```

(3)
```
using System;
namespace ex05_03
{
    class TCdays
    {
        string[] days = { "星期日", "星期一", "星期二", "星期三", "星期四", "星期五",
                          "星期六" };
        public string this[int index]
        {
```

```csharp
                get
                {
                    if (index >= 0 && index <= 6)
                        return days[index];
                    else
                        return days[0];
                }
            }
        }
        class TEdays
        {
            string[] days = { "Sunday", "Monday", "Tuesday", "Wednesday", "Thursday",
                "Friday", "Satday" };
            public string this[int index]
            {
                get
                {
                    if (index >= 0 && index <= 6)
                        return days[index];
                    else
                        return days[0];
                }
            }
            private int getday(string testDay)
            {
                int i = 0;
                foreach( string day in days)
                {
                    if (day == testDay)
                        return i;
                    i++;
                }
                return -1;
            }
            public int this[string day]
            {
                get
                {
                    return(getday(day));
                }
            }
        }
        class Program
        {
            static void Main(string[] args)
            {
                int i;
                string s;
                TCdays cday = new TCdays();
                TEdays eday = new TEdays();
                Console.WriteLine("请输入星期几(0-6分别代表星期日到星期六)");
```

```csharp
                i = int.Parse(Console.ReadLine());
                Console.WriteLine("{0}-----{1}", cday[i],eday[i]);        //输出_____
                Console.WriteLine("请输入星期几的英文表示：:");
                s = Console.ReadLine();
                Console.WriteLine("{0}-----{1}",eday[eday[s]],eday[s]);   //输出_____
            }
        }
    }
```

（4）
```csharp
using System;
namespace ex05_04
{
    class Myobj
    {
        private int num;
        public Myobj() { }
        public Myobj(int n)
            { num = n; }
        public static Myobj operator +(Myobj obj1,Myobj obj2)
        {
            return new Myobj(obj1.num + obj2.num);
        }
        public void disp()
        {
            Console.WriteLine("num = {0}", num);
        }
    }
    class Program
    {
        static void Main(string[] args)
        {
            Myobj obj1 = new Myobj(5), obj2 = new Myobj(8), obj3;
            obj3 = obj1 + obj2;
            obj3.disp();                         //输出_____
        }
    }
}
```

2. 编写一个程序，采用一个类求 n!，并分别求出 3!，6!，10! 及 15! 的值。请完善程序。

```csharp
using System;
namespace ex05_05
{
    class fac
    {
        int n0;                             //待求的阶乘
        uint p;                             //存储 n0!
        int overflow;                       //溢出标志
        _____①_____ ;                //默认形式构造函数,给n0赋值为1
        public fac(int j)
```

```
        {
            n0 = j;
            overflow = 0;                    //溢出标志
            p = 1;
            uint i, p1 = p;
            if (j >= 0)
                for (i = 1; i <= j; i++)
                {
                    p1 = p * i;
                    if (p != p1 / i)         //若 i!正确,则有(i-1)!等于 i!/i
                    {
                        Console.WriteLine("{0}!数据溢出了", i);
                        overflow = 1;
                        break;
                    }
                    p = p1;
                }
            else
                Console.WriteLine("数据错误");
        }
        public void display()
        {
            if (    ②    )
                Console.WriteLine("{0}!= {1}",n0, p);
            else
                Console.WriteLine("{0}太大", n0);
        }
    };
    class Program
    {
        static void Main(string[] args)
        {
            int n;
            Console.WriteLine("请输入一个整数");
            n = int.Parse(Console.ReadLine());
            fac a = new fac(n), b;
            b = new fac();
            a.display();
            b.display();
        }
    }
}
```

3. 定义盒子 Box 类,要求用多个方法分别实现:设置盒子各棱长,计算盒子体积、盒子表面积,输出盒子棱长、体积、表面积。请修改程序中的错误,并上机运行。

```
using System;
namespace ex05_06
{
    class Box
    {
```

```csharp
        int x, y, z, v, s;
        public void init( int x1 = 0, int y1 = 0, int z1 = 0)
        { x = x1; y = y1; z = z1; }
        void volume() { v = x * y * z; }
        void area() { s = 2 * (x * y + x * z + y * z); }
        public void display()
        {
            Console.WriteLine("x = {0},y = {1},z = {2}", x, y, z);
            Console.WriteLine("s = {0},v = {1}",s,v);
        }
    };
    class Program
    {
        static void Main(string[ ] args)
        {
            Box a ;
            a.init(2, 3, 4);
            a.volume();
            a.area();
            a.display();
        }
    }
}
```

4. 设计一个复数类 Complex，实现两个复数的乘法运算。Complex 类包括复数的实部和虚部，以及实现复数相乘的方法 mult()和输出复数的方法 display()。请完善程序。

```csharp
using System;
namespace ex05_07
{
    class Complex
    {
        float a, b;                                //分别表示实部和虚部
        public Complex() { }
        public Complex(float x, float y)
        { a = x; b = y; }
        public Complex mult(ref Complex s)          //对象引用作为参数
        {
            float x = a * s.a - b * s.b;
            float y =    ①     ;
            a = x; b = y;
            return this;
        }
        public void display()
        {
            if (b > 0)
                Console.WriteLine("{0}+{1}i", a, b);
            else
                Console.WriteLine("{0}-{1}i", a, -b);
        }
    };
```

```
class Program
{
    static void Main(string[] args)
    {
        Complex s1 = new Complex(2,3),s2 = new Complex(3, 4);
        Console.WriteLine("复数 s1:");
        s1.display();
        Console.WriteLine("复数 s1:");
        s2.display();
        _____②_____ ;                          //s1 = s1 × s2
        Console.WriteLine("复数 s1Xs2 = :");
        s1.display();
    }
}
```

三、独立编程实验

1. 定义一个长方体类,其数据成员包括 length、width、height,分别代表长方体的长、宽、高。要求用方法实现以下功能:

(1) 由键盘输入长方体的长、宽、高。

(2) 计算长方体的体积。

(3) 输出长方体的体积。

(4) 编写主函数使用这个类。

2. 定义一个学生类,其中有 3 个数据成员(学号、姓名、年龄)以及若干方法。要求:

(1) 用方法实现对数据的输入输出。

(2) 用构造函数初始化数据成员,使用析构函数输出数据成员。

3. 设计一个素数类 Prime,要求能够求任意区间的全部素数以及这些素数之和,并利用该类中的方法求[2,1000]中的全部素数之和。

4. 设计一个航班类 Flights,具有机型、班次、额定载客数和实际载客数等数据成员,还具有输入输出数据成员以及求载客效率(载客效率=实际载客数/额定载客数)的功能。

5. 建立一个分数类。分数类的数据成员包括分子和分母,方法包括约分、通分、加、减、乘、除、求倒数、比较、显示和输入。

分数类声明形式如下:

```
class fraction
{
    int above;                                       //分子
    int below;                                       //分母
    void reduction() { }                             //约分
    void makecommond(ref fraction s) { }             //通分
    public fraction(int x = 0, int y = 1)
    { above = x;below = y; }
    public void add(ref fraction s) { }              //分数加法
    publicvoid sub(ref fraction s) { }
    publicvoid mul(ref fraction s){ }
```

```
publicvoid div(ref fraction s) { }
publicvoid reciprocal() { }                         //求倒数
public bool equal(ref fraction s) {}                //等于运算
public bool greaterthan(ref fraction s)    {    }   //大于运算
public bool lessthan(ref fraction s)       {    }   //小于运算
public void display() { }                           //显示
void input() { }                                    //输入
};
```

实验6 面向对象编程(二)

一、实验目的

1. 熟悉继承的概念和设计方法。
2. 掌握多态的概念和设计方法。
3. 掌握接口的概念和设计方法。
4. 掌握委托和事件的概念和设计方法。
5. 进一步掌握面向对象程序设计的方法。

二、模仿编程实验

1. 阅读程序,写出输出结果并上机验证。

(1)
```
using System;
namespace ex06_01
{
    class BaseA
    {
        private int n;
        protected int m;
        public BaseA( int x, int y)
        {
            n = x;
            m = y;
            Console.WriteLine("In class BaseA Constructor function");
        }
        public void funa()
        {
            Console.WriteLine("class BaseA:n = {0},m = {1}",n,m);
        }
    }
    class DerivedB : BaseA
    {
        public DerivedB( int x, int y, int z) : base(x, y)
        {
            Console.WriteLine("In class DerivedB Constructor function");
        }
```

```csharp
    class Program
    {
        static void Main(string[] args)
        {
            DerivedB obj1 = new DerivedB(1, 2, 3);
            obj1.funa();
        }
    }
}
```

(2)

```csharp
using System;
namespace ex06_02
{
    class B1
    {
        protected int d1;
        public B1(int a = 0)
        {
            d1 = a;
            Console.WriteLine("In B1 Constructor");
        }
        ~B1()
        {
            Console.WriteLine("In B1 Destructor");
        }
    }
    class B2 : B1
    {
        protected int d2;
        public B2(int a = 0) : base(a)
        {
            d2 = a;
            Console.WriteLine("In B2 Constructor");
        }
        ~B2()
        {
            Console.WriteLine("In B2 Destructor");
        }
    }
    class C : B2
    {
        int d;
        public C(int x, int z) : base(x)
        {
            d = z;
            Console.WriteLine("In Constructor C");
        }
        ~C()
        {
            Console.WriteLine("In C Destructor");
        }
```

```csharp
            public void Show()
            {
                Console.WriteLine("{0},{1},{2}", d1, d2, d);
            }
        }
        class Program
        {
            static void Main(string[] args)
            {
                C obj = new C(5, 9);
                obj.Show();
            }
        }
    }
```

(3)
```csharp
using System;
namespace ex06_03
{
    class Base
    {
        public void fun1()
        {
            Console.WriteLine("Base 类中的 fun1 方法");
        }
        public virtual void fun2()
        {
            Console.WriteLine("Base 类中的 fun2 方法");
        }
    }
    class Derived:Base
    {
        new public void fun1()
        {
            Console.WriteLine("Derived 类中的 fun1 方法");
        }
        public override void fun2()
        {
            Console.WriteLine("Derived 类中的 fun2 方法");
        }
    }
    class Program
    {
        static void Main(string[] args)
        {
            Derived objD = new Derived();
            Base objB = objD;
            objB.fun1();
            objD.fun1();
            objB.fun2();
```

```
            objD.fun2();
        }
    }
}
```

2. 先定义点类 Point，再由点类 Point 派生出圆类 Circle，主程序输出圆心坐标和面积。请完善程序。

```
using System;
namespace ex06_04
{
    class Point
    {
        int x, y;                                    //圆心坐标
        publicPoint(int a = 0, int b = 0)
        {
            x = a;
            y = b;
        }
        ~Point() { }
        public int getx
        {
            get { return x; }
        }
        public int gety
        {
            get { return y; }
        }
    };
    class Circle :      ①
    {
        int r;                                       //半径
        public Circle(int x, int y, int ra) :base(x, y)
        {
            r = ra;
        }
        public int getr
        {
            get { return r; }
        }
        ~Circle() { }
        public double area()
        {
            return 3.14 * r * r;
        }
    };
    class Program
    {
        static void Main(string[] args)
        {
            Circle c1 = new Circle(100, 300, 10);
```

```
            Console.WriteLine("圆心是({0},{1})",_____②_____,_____③_____);
            Console.WriteLine("半径为{0}的圆面积是{1}",_____④_____,_____⑤_____);
        }
    }
}
```

3. 下面的程序在一个类中实现了多个接口，请按要求完善程序。

```
using System;
namespace ex06_05
{
    public interface shape
    {
        double area();                          //接口方法,计算图形面积
    }
    public interface display
    {
        _____①_____;                          //接口方法,显示图形面积
    }
    public class Circle:_____②_____
    {
        private double radius;
        public Circle(double x)
        {
            radius = x;
        }
        public double area()
        {
            return (3.14 * radius * radius);
        }
        public void displayresult()
        {
            Console.WriteLine("半径为{0}的圆面积为{1}", radius, area());
        }
    }
    class Program
    {
        static void Main(string[] args)
        {
            Circle cc1 = new Circle(10);
            cc1.displayresult();
            _____③_____;                      //实例化一个半径为5的圆
            cc1.displayresult()
        }
    }
}
```

4. 改正下面程序中的错误，并写出程序运行的结果。

```
using System;
namespace ex06_06
{
```

```csharp
public class Wmessage
{
    private string wname;
    public delegate void mydelegateType();
    public event mydelegateType MyClickEvent;
    public Wmessage(string name)
    {
        this.wname = name;
    }
    public void Click()
    {
        Console.WriteLine(wname + "被点击了一下");
        if (MyClickEvent != null)
            MyClickEvent();
    }
}
public class Cwindows
{
    private string cname;
    public Cwindows(string name)                  //构造函数,给窗口取名
    { this.cname = name; }
    public void Max()
    {
        Console.WriteLine(cname + "窗口最大化");
    }
    public void Min()
    {
        Console.WriteLine(cname + "窗口最小化");
    }
}
class Program
{
    static void Main(string[] args)
    {
        Wmessage m = new Wmessage("鼠标左键");
        Cwindows c1 = new Cwindows(), c2;      //实例化窗口 c1 和 c2
        m.MyClickEvent += new Wmessage.mydelegateType(c1.Max);
        m.MyClickEvent += new Wmessage.mydelegateType(c2.Min);
        m.Click();
    }
}
```

三、独立编程实验

1. 设计一个汽车类 vehicle,包含的数据成员有车轮个数 wheels 和车重 weight。小车类 car 是 vehicle 的私有派生类,其中包含载人数 passenger_load;卡车类 truck 是 vehicle 的私有派生类,其中包含载人数 passenger_load 和载重量 payload。每个类都有相关数据的输出方法。

提示：vehicle 类是基类，由它派生出 car 类和 truck 类，将公共的属性和方法放在 vehicle 类中，并编写 Main() 方法构成完整程序。

2. 设计一个圆接口 circle（包含一个求面积方法）和一个桌子接口 table（包含一个输出方法）；另设计一个圆桌类 roundtable，它是从前两个接口派生的，要求输出一个圆桌的高度、面积和颜色等数据。编写 Main() 方法构成完整程序。

3. 设计一个程序，通过委托方式求两个整数（x, y）的立方和（x^3+y^3）及立方差（x^3-y^3），并编写 Main() 方法构成完整程序。

4. 定义一个字符串类 onestr，包含一个存放字符串的成员变量，能够通过构造函数初始化字符串，通过成员方法显示字符串的内容。在此基础上派生出 twostr 类，增加一个存放字符串的成员变量，并能通过派生类的构造函数传递参数，初始化两个字符串，通过另一成员函数进行两个字符串的合并以及输出，并编写 Main() 方法构成完整程序。

实验 7 复杂数据表示与应用

一、实验目的

1. 掌握数组的定义和使用方法。
2. 掌握交错数组的定义和使用方法。
3. 掌握与数组有关的常用算法。
4. 掌握枚举的定义和使用。
5. 掌握结构的定义和使用。

二、模仿编程实验

1. 阅读以下程序，写出它们的输出结果并上机验证。可单步跟踪调试程序，观察数组元素值如何变化。

(1)
```
using System;
namespace ex07_01
{
    class Program
    {
        static void Main(string[] args)
        {
            int []num = new int[10] { 1,0,0,0,0,0,0,0,0,0 };
            int i, j;
            for (j = 0; j < 10; ++j)
                for (i = 0; i < j; ++i)
                    num[j] = num[j] + num[i];
            for (j = 0; j < 10; ++j)
                Console.WriteLine("{0}", num[j]);
        }
    }
}
```

（2）

```
using System;
namespace ex07_02
{
    class Program
    {
        static void Main(string[] args)
        {
            int i, j;
            const int N = 7;
            int[] a = new int[N]{ 5, 3, 4, 7, 3, 5, 6 }, b = new int[N];
            for (i = 0; i < N; i++)
                Console.Write("{0},", a[i]);
            Console.WriteLine();
            for (i = 0; i < N; i++)
                for (j = 0, b[i] = 0; j < N; j++)
                    if (a[j] < a[i]) b[i]++;
            for (i = 0; i < N; i++)
                Console.Write("{0},", b[i]);
            Console.WriteLine();
        }
    }
}
```

（3）

```
using System;
namespace ex07_03
{
    class Program
    {
        static void Main(string[] args)
        {
            int s = 0;
            int[][] a = new int[2][];
            a[0] = new int[4] { 1, 9, 3, 6 };
            a[1] = new int[3] { 4, 5, 7 };
            for (int i = 0; i < a.Length; i++)
                for (int j = 0; j < a[i].Length; j++)
                    s += a[i][j];
            Console.WriteLine(s);
        }
    }
}
```

2. 已知数列

$$\begin{cases} f_1 = 1, f_2 = 1, f_3 = 1 \\ f_n = f_{n-1} - 2f_{n-2} + f_{n-3} \quad (n > 3) \end{cases}$$

按照每行 5 个元素的格式输出 $f_1 \sim f_{20}$，并输出其中最大、第二大两项的值。请完善程序。

```csharp
using System;
namespace ex07_04
{
    class Program
    {
        static void Main(string[] args)
        {
            int i; int [] f = new int [21] ;
            int m1, m2;
            ___①___ ;
            for (i = 4; i < 21; i++)
                f[i] = f[i - 1] - 2 * f[i - 2] + f[i - 3];
            for (i = 0; i < 21; i++)
            {
                if (i % 5 == 0)
                    ___②___ ;
                Console.Write("\t{0}", f[i]);
            }
            Console.WriteLine();
            m1 = m2 = f[1];
            for (i = 2; i < 21; i++)
                if (m1 < f[i])
                {
                    ___③___ ;
                    m1 = f[i];
                }
            Console.WriteLine("\n最大项的值:{0},次大项的值:{1}", m1, m2);
        }
    }
}
```

3. 已知下面程序先计算出杨辉三角数表存于二维数组 a 的对应位置各元素中，然后输出数组 a 主对角元素及以下的元素便得到如下杨辉三角数表的前几行。

```
1
1   1
1   2   1
1   3   3   1
1   4   6   4   1
```

请补充完善程序。

```csharp
using System;
namespace ex07_05
{
    class Program
    {
        static void Main(string[] args)
        {
            int i, j;
```

```
            int[ , ]a = new int[5,5];
            for(i = 0;i < 5;i++)                    //求三角数表
            {
                a[i,0] = 1;
                _____①_____ ;
                for(j = 1;_____②_____;j++)
                    a[i,j] = a[i - 1,j - 1] + a[i - 1,j];
            }
            for(i = 0;i < 5;i++)                    //输出三角数表
            {
                Console.WriteLine();
                for (j = 0;_____③_____ ; j++)
                    Console.Write("\t{0}", a[i, j]);
            }
            Console.WriteLine();
        }
    }
}
```

4. 下面程序计算两个复数之积,复数用结构表示。请完善程序。

```
using System;
namespace ex07_06
{
    public struct complex
    {
        public double re, im;
    };
    class c
    {
        public static complex mul(complex z1, complex z2)
        {
            complex z = new complex();
            z.re = _____①_____ ;
            z.im = _____②_____ ;
            return z;
        }
        static void print(complex z)
        {
            if (z.im < 0)
                Console.WriteLine("{0} - {1}", z.re, - z.im);
            else
                Console.WriteLine("{0} + {1}", z.re, z.im);
        }
    }
    class Program
    {
        static void Main(string[ ] args)
        {
            complex z1 = new complex(), z2 = new complex(), z3 = new complex();
            Console.WriteLine("复数 z1 实部: ");
```

```
            z1.re = int.Parse(Console.ReadLine());
            Console.WriteLine("复数 z1 虚部: ");
            z1.im = int.Parse(Console.ReadLine());
            Console.WriteLine("复数 z2 实部: ");
            z2.re = int.Parse(Console.ReadLine());
            Console.WriteLine("复数 z2 虚部: ");
            z2.im = int.Parse(Console.ReadLine());
            z3 = c.mul(z1, z2);
            Console.WriteLine("z1 * z2 = {0} + {1}i", z3.re, z3.im);
        }
    }
}
```

三、独立编程实验

1. 从键盘输入某班学生程序设计课程考试成绩,评定每个学生的成绩等级。如果高于平均分 10 分,则等级为"优秀";如果低于平均分 10 分,则等级为"一般";否则等级为"中等"。

2. 编写静态函数 public static int lookup(int[] a,int y),在数组 a 中查找是否有等于 y 的元素(a.length 可以返回数组的元素个数),若有,返回第一个相等元素的下标,否则,返回 -1。编写主函数调用它。

3. 编写静态函数 public static int maxf(int [] x),然后调用该函数返回 20 个数中的最大数。编写主函数调用它,输出 20 个数中的最大数。

4. 输入 4×4 的数组,求:
 (1) 对角线上行、列下标均为偶数的各元素的积。
 (2) 找出对角线上其值最大的元素和它在数组中的位置。

5. 有 n 名学生的数据,每个学生的数据包括学号、姓名、性别、三门课的考试成绩及平均成绩。学生信息结构及 Main 方法如下:

```
public struct student
{
    public string num;           //学号
    public string name;          //姓名
    public string sex;           //性别
    public int[] score;          //score[1]~score[3]存放三门课成绩,score[0]存放平均成绩
};
class Program
{
    static void Main(string[] args)
    {
        student[] s = new student[5];
        int i;
        for(i = 0;i < s.Length;i++)
            s[i].score = new int[4];
        p.input(ref s);
        p.sort(ref s);
        p.output(s);
```

　　　　}
　　}

（1）编写一个 public static void input(ref student[] s)函数，用来输入 n 个学生的信息，并计算每人的平均成绩。

（2）编写一个 public static void output(student[] s)函数，用来输出 n 个学生的信息。

（3）编写一个 public static void sort(ref student[] s)函数，对 n 个学生的成绩按平均成绩由小到大进行排序。

实验 8　Windows 窗体与控件

一、实验目的

1. 掌握 C#窗体的属性及设计方法。
2. 掌握 C#中常见控件的属性、方法。
3. 掌握 C#中窗体及常见控件的事件处理机制和设计方法。
4. 掌握 C#中使用各种常用控件设计窗体界面的方法。

二、模仿编程实验

1. 在窗口中放置一个标签控件，单击该标签生成如图 2-1 所示的信息。请完善程序。

```
using System;
using System.Windows.Forms;
namespace ex08_01
{
    public partial class Form1 : Form
    {
        public Form1()
        {
            InitializeComponent();
        }
        struct Student
        {
            public int no;
            public string name;
            public char sex;
            public int score;
            public string getinfo()
            {
                string result = "该学生的信息是：";
                result += "\n 学号：" + no;
                result += "\n 姓名：" + name;
                result += "\n 性别：" + sex;
                result += "\n 成绩：" + score;
                ____①____ ;
            }
        }
```

图 2-1　标签应用示例

```csharp
        private void label1_Click(object sender, EventArgs e)
        {
            Student stu;
            stu.no = 1001;
            stu.name = "张三";
            stu.sex = '男';
            stu.score = 623;
            Mylabel.Text = ②;
            Mylabel.Text += "\n\n" + DateTime.Now;
        }
    }
}
```

2. 设计一个应用程序,界面如图 2-2 所示,单击单选按钮选择答案,单击"确定"按钮后根据所选答案给出相应的信息。请完善程序。

图 2-2 单选按钮示例

```csharp
using System;
using System.Windows.Forms;
namespace ex08_02
{
    public partial class Form1 : Form
    {
        public Form1()
        {
            InitializeComponent();
        }
        private void button1_Click(object sender, EventArgs e)
        {
            if (radioButton3.____①____)
                MessageBox.Show("恭喜您答对了!", "信息提示", MessageBoxButtons.OK);
            else if (radioButton1.____②____)
                MessageBox.Show("答错了,数据库管理系统不属于应用软件",
                                "信息提示", MessageBoxButtons.OK);
            else if (radioButton2.____③____)
                MessageBox.Show("答错了,数据库管理系统不是存储器",
                                "信息提示", MessageBoxButtons.OK);
            else
                MessageBox.Show("答错了,数据库管理系统不是计算机系统",
                                "信息提示", MessageBoxButtons.OK);
```

        }
    }
}

3. 设计一个应用程序，界面如图 2-3 所示，输入相关信息，单击"确定"按钮后输出相应的信息，如图 2-4 所示。请完善程序。

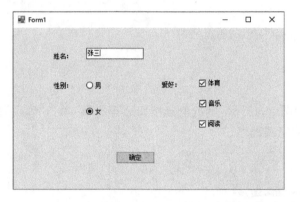

图 2-3 单选按钮和复选框的应用程序界面

图 2-4 信息提示窗口

```
using System;
using System.Windows.Forms;
namespace ex08_03
{
    public partial class Form1 : Form
    {
        public Form1()
        {
            InitializeComponent();
        }
        private void button1_Click(object sender, EventArgs e)
        {
            string s = "";
            s = s + textBox1.Text + " 性别:";
            if (radioButton1.Checked)
                s = s + radioButton1.Text;
            else
                s = s + radioButton2.Text;
            s = s + "\n爱好:";
            if (      ①      )
                s = s + checkBox1.Text;
            if (checkBox2.Checked)
                s = s + "和" +      ②      ;
            if (checkBox3.Checked)
                s = s + "和" +      ③      ;
            MessageBox.Show(s, "信息提示", MessageBoxButtons.OK);
        }
    }
}
```

4. 设计一个窗体应用程序，其界面如图 2-5 所示，单击其中的按钮可在两个列表框中移动数据项。请完善程序。

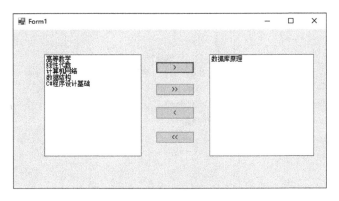

图 2-5　列表框的应用程序界面

```
using System;
using System.Windows.Forms;
namespace ex08_04
{
    public partial class Form1 : Form
    {
        public Form1()
        {
            InitializeComponent();
        }
        private void Form1_Load(object sender, EventArgs e)
        {
            listBox1.Items.Add("高等数学");
            listBox1.Items.Add("线性代数");
            listBox1.Items.Add("数据库原理");
            listBox1.Items.Add("计算机网络");
            listBox1.Items.Add("数据结构");
            listBox1.Items.Add("C#程序设计基础");
            checklist();
        }
        private void checklist()
        {
            if (listBox1.Items.Count == 0)
                button1.Enabled = false;     ①     ;
            else
                button1.Enabled = true;      ②     ;
            if (     ③     )
                button3.Enabled = false;button4.Enabled = false;
            else
                button3.Enabled = true;button4.Enabled = true;
        }
        private void button1_Click(object sender, EventArgs e)
        {
            if (listBox1.SelectedIndex >= 0)
```

```
            {
                listBox2.Items.Add(listBox1.SelectedItem);
                listBox1.Items.Remove(listBox1.SelectedItem);
            }
            checklist();
        }
        private void button2_Click(object sender, EventArgs e)
        {
            foreach(object item in listBox1.Items)
                listBox2.Items.Add(item);
            listBox1.Items.Clear();
            checklist();
        }
        private void button3_Click(object sender, EventArgs e)
        {
            if (listBox2.SelectedIndex >= 0)
            {
                    ④    ;
                listBox2.Items.Remove(listBox2.SelectedItem);
            }
            checklist();
        }
        private void button4_Click(object sender, EventArgs e)
        {
            foreach (object item in listBox2.Items)
                listBox1.Items.Add(item);
            listBox2.Items.Clear();
            checklist();
        }
    }
}
```

三、独立编程实验

1. 编写一个程序,计算两个指定年份之间的闰年并输出,运行界面如图 2-6 所示。

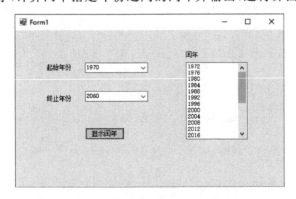

图 2-6　下拉列表框的应用程序运行界面

2. 编写一个简单的计算器,能够实现整数的加减乘除 4 种运算,设计界面如图 2-7 所示。

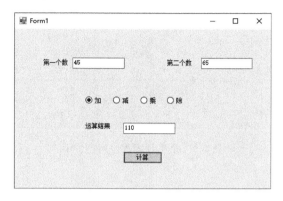

图 2-7　计算器程序运行界面

3. 编写程序,根据不同的选项,显示不同的图片,设计界面如图 2-8 所示。不同类型的图片可事先从网上查找下载到本地。

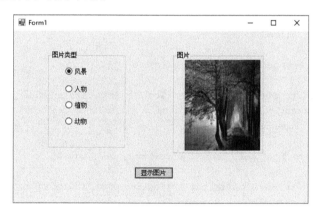

图 2-8　图片显示程序运行界面

实验 9　用户界面设计

一、实验目的

1. 掌握 C# 下拉式菜单的设计方法。
2. 掌握 C# 弹出式菜单的设计方法。
3. 掌握通用对话框控件的设计方法。
4. 掌握 ImageList、TreeView 和 ListView 等控件的设计方法。

二、模仿编程实验

1. 创建如图 2-9 所示的应用程序菜单,并可通过按钮设置菜单项是否可用。请完善程序。

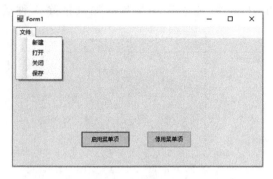

图 2-9 菜单设置程序运行界面

```
using System;
using System.Windows.Forms;
namespace ex09_01
{
    public partial class Form1 : Form
    {
        public Form1()
        {
            InitializeComponent();
        }
        private void button1_Click(object sender, EventArgs e)
        {
            ToolStripMenuItem myitem = (ToolStripMenuItem)menuStrip1.Items[0];
            foreach(ToolStripMenuItem item in myitem.DropDownItems)
                item.Enabled = true;
        }
        private void button2_Click(object sender, EventArgs e)
        {
            ToolStripMenuItem myitem = (ToolStripMenuItem)menuStrip1.Items[0];
            foreach (     ①     )
                    ②    ;
        }
    }
}
```

2. 在程序运行过程中动态创建如图 2-10 所示的级联菜单,并添加相应的单击事件代码。请完善程序。

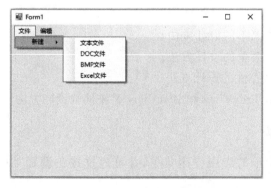

图 2-10 菜单操作程序运行界面

```csharp
using System;
using System.Windows.Forms;
namespace ex09_02
{
    public partial class Form1 : Form
    {
        public Form1()
        {
            InitializeComponent();
        }
        private void Form1_Load(object sender, EventArgs e)
        {
            ToolStripMenuItem F_ts = (ToolStripMenuItem)menuStrip1.Items[0];
            ToolStripMenuItem ts1 = new ToolStripMenuItem("文本文件");
            ToolStripMenuItem ts2 = new ToolStripMenuItem("DOC 文件");
            ToolStripMenuItem ts3 = new ToolStripMenuItem("BMP 文件");
            ToolStripMenuItem ts4 = new ToolStripMenuItem("Excel 文件");
            ToolStripMenuItem F_ts1 = (ToolStripMenuItem)F_ts.DropDownItems[0];
            F_ts1.DropDownItems.Add(ts1);
            F_ts1.DropDownItems.Add(ts2);
            _____①_____ ;
            F_ts1.DropDownItems.Add(ts4);
            ts1.Click += new EventHandler(ts1click);
            ts2.Click += new EventHandler(ts2click);
            _____②_____ ;
            ts4.Click += new EventHandler(ts4click);
        }
        private void ts1click(object sender,EventArgs e)
        {
            MessageBox.Show("新建文本文件");
        }
        private void ts2click(object sender, EventArgs e)
        {
            MessageBox.Show("新建 DOC 文件");
        }
        private void ts3click(object sender, EventArgs e)
        {
            MessageBox.Show("新建 BMP 文件");
        }
        private void ts4click(object sender, EventArgs e)
        {
            MessageBox.Show("新建 Excel 文件");
        }
    }
}
```

3. 创建如图 2-11 所示的应用程序,添加工具栏和文件打开对话框组件,在工具栏里设置几个图像按钮,分别用于打开文本文件和图像文件,打开的文件和图像分别显示在窗口中的 TextBox 和 PixtureBox 里。请完善程序。

图 2-11 图像操作程序运行界面

```
using System;
using System.Drawing;
using System.Text;
using System.Windows.Forms;
namespace ex09_03
{
    public partial class Form1 : Form
    {
        public Form1()
        {
            InitializeComponent();
        }
        private void toolStripButton3_Click(object sender, EventArgs e)
        {
            openFileDialog1.Filter = "txt 文件(*.txt)|*.txt";
            if (openFileDialog1.ShowDialog() == DialogResult.OK)
            {
                System.IO.StreamReader sr = new
                System.IO.StreamReader(openFileDialog1.FileName, Encoding.Default);
                textBox1.Text = sr.ReadToEnd();
                sr.Close();
            }
        }
        private void toolStripButton4_Click(object sender, EventArgs e)
        {
            openFileDialog1.Filter = "*.jpg|*.jpg|*.bmp|*.bmp";
            if (    ①    )
                pictureBox1.Image = Image.FromFile(    ②    );
        }
    }
}
```

三、独立编程实验

1. 设计一个下拉式菜单实现对用户输入的两个数的加、减、乘、除运算,程序界面如图 2-12 所示。

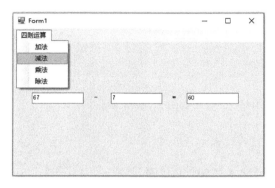

图 2-12　四则运算程序运行界面

2. 设计一个多文档窗口应用程序,在"题目"菜单下选择"加""减""乘""除"子菜单可以分别打开子窗口,给出两位随机整数的加减乘除题目,在"窗口"菜单下,可对已出现的子窗口进行不同的排列,在子窗口中通过"评判"按钮给出对错,程序运行界面如图 2-13 所示。

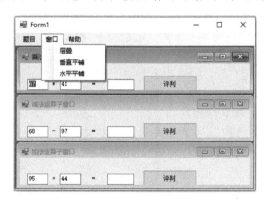

图 2-13　多文档程序运行界面

下面是主窗口 Form1 的部分代码:

```
private void 加ToolStripMenuItem_Click(object sender, EventArgs e)
{
    Form2 child1 = new Form2();
    child1.MdiParent = this;
    child1.Show();
    child1.Text = "加法运算子窗口";
    child1.mylabel1.Text = " + ";
}
```

下面是子窗口 Form2 的部分代码:

```
private void Form2_Load(object sender, EventArgs e)
{
```

```
            Random ran = new Random();
            textBox1.Text = ran.Next(10, 99).ToString();
            textBox2.Text = ran.Next(10, 99).ToString();
        }
        public Label mylabel1
        {
            get
            {
                return label1;
            }
        }
        private void button1_Click(object sender, EventArgs e)
        {
            if (label1.Text == " + ")
            {
                if (int.Parse(this.textBox1.Text) + int.Parse(this.textBox2.Text)
                    == int.Parse(this.textBox3.Text))
                    MessageBox.Show("right");
                else
                    MessageBox.Show("error");
            }
        }
```

实验 10 文 件 操 作

一、实验目的

1. 掌握 C#中文件的类型和特点。
2. 掌握 C#中文本文件和二进制文件的读写方法。
3. 掌握 C#中序列化和反序列化存取文件的方法。

二、模仿编程实验

1. 求[3,100]中的素数,并将它们输出到屏幕和名为 sushu.txt 的文本文件中,数据之间用 Tab 键隔开,每输出 5 个素数则自动换行。

参考程序如下,请完善程序后上机运行,并用记事本打开程序生成的 sushu.txt 文件,检验结果是否正确。

```
using System;
using System.IO;
namespace ex10_01
{
    class Program
    {
        static void Main(string[] args)
        {
            int flag,n = 0;
            StreamWriter outfile = _____①_____;        //定义文件输出流,并打开
```

```
            Console.WriteLine("3~100 之间的素数是：");
            for (int m = 3; m < 100; m += 2)
            {
                int k = (int)(Math.Sqrt(m));
                flag = 1;
                for (int i = 2; i <= k; i++)
                    if (m % i == 0)
                    {
                        flag = 0;
                        break;
                    }
                if (flag == 1)
                {
                    Console.Write("{0}\t", m);
                    ____②____ ;                    //输出数据到文件
                    n = n + 1;
                    if (n % 5 == 0)
                    {
                        Console.Write("\n");
                        outfile.Write("\n");
                    }
                }
            }
            ____③____ ;                            //关闭文件
        }
    }
}
```

2. 编写程序统计上述 sushu.txt 文本文件中素数的个数。请完善程序。

```
using System;
using System.IO;
namespace ex10_02
{
    class Program
    {
        static void Main(string[] args)
        {
            int x, y = 0, n = 0;
            try {
                StreamReader infile = ____①____ ;   //定义文件输入流对象,并打开源文件
                while ( ____②____ )                 //读数据,并判断是否读到文件尾部
                {
                    if (x >= '0' && x <= '9')
                        y = y * 10 + (x - '0');
                    else
                    {
                        if (y > 0)
                        {
                            Console.Write("{0} ", y);
                            ____③____ ;             //计数
```

```
                                y = 0;
                            }
                        }
                    }
                    Console.WriteLine("素数个数是{0}", n);
                        ____④____ ;                           //关闭文件
                }
                catch
                { }
            }
        }
    }
```

3. 从键盘输入若干条记录,每条记录包含姓名、年龄、工资这三个数据项,将信息存入二进制文件 gongzi.dat。请完善程序。

```
using System.IO;
using System.Text;
namespace ex10_03
{
    class Program
    {
        struct Person
        {
            public int pno;
            public string pname;
            public double salary;
        }
        static void Main(string[] args)
        {
            string path = "e:\\gongzi.dat";
            const int pnum = 4;
            Person[] persons = new Person[pnum];
            persons[0].pno = 1;persons[0].pname = "张三";persons[0].salary = 4500;
            persons[1].pno = 2;persons[1].pname = "李四";persons[1].salary = 4600;
            persons[2].pno = 3;persons[2].pname = "王五";persons[2].salary = 5100;
            persons[3].pno = 4;persons[3].pname = "刘齐";persons[3].salary = 3900;
            int i;
            if (File.Exists(path))
                File.Delete(path);                    //存在该文件时删除它
            FileStream outfile = File.OpenWrite(____①____);
            BinaryWriter bf = new BinaryWriter(____②____, Encoding.Default);
            for (i = 0;i < pnum;i++)
            {
                bf.Write(persons[i].pno);
                bf.Write(persons[i].pname);
                bf.Write(persons[i].salary);
            }
            outfile.close();
                ____③____ ;
        }
    }
}
```

4. 下面的程序功能是读出二进制文件"E:\gongzi.dat"并在屏幕上显示出来。请完善程序。

```
using System;
using System.IO;
using System.Text;
namespace ex10_04
{
    class Program
    {
        static void Main(string[] args)
        {
            string path = "E:\\gongzi.dat";
            string fstr = "";
            FileStream infile = File.OpenRead(path);
            BinaryReader fb = new BinaryReader(infile, Encoding.Default);
            infile.Seek(0, SeekOrigin.Begin);
            while (fb.PeekChar() > -1)
                fstr = fstr + fb.ReadInt32() + "\t" + fb.ReadString()
                    + "\t" + fb.ReadDouble() + "\n";
            _____①_____ ;
            fb.Close();
            _____②_____ ;
        }
    }
}
```

三、独立编程实验

1. 编写程序实现将文本文件 s.txt 的内容复制到另一文件 d.txt 中。

2. 一个数 n 是完数，即 n 的真因子之和等于该数本身，如 6=1+2+3，则 6 是完数。求出 1~1000 范围内的完数，将其写入二进制文件 wnum.dat 中，并打开该文件，读出所有的完数并显示出来。

3. 利用序列化和反序列化编写程序实现：
(1) 建立一个通讯簿文件，存入三个联系人的数据（包括姓名、家庭电话、手机号）。
(2) 从键盘上输入另外一个联系人的数据，增加到文件的末尾。
(3) 输出文件中的全部数据。
(4) 从键盘输入一个手机号码，查询文件中有无此手机号，如有，则显示该联系人全部信息。如没有，则显示"无此电话！"。

实验 11　图形与图像处理

一、实验目的

1. 掌握 C♯ 中绘制图形的基本方法。
2. 掌握 C♯ 中各种绘图工具的用法。

3. 掌握 C#中绘制文字的方法。
4. 掌握 C#中图像打开、保存、显示的方法。
5. 掌握 C#中常用的图像处理方法。

二、模仿编程实验

1. 设计一个图形用户界面的绘图程序，界面如图 2-14 所示，单击按钮后绘制相应的直线。请完善程序。

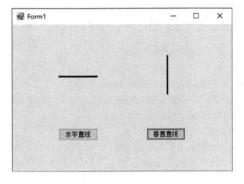

图 2-14 绘制直线程序运行界面

```
using System;
using System.Drawing;
using System.Windows.Forms;
namespace ex11_01
{
    public partial class Form1 : Form
    {
        public Form1()
        {
            InitializeComponent();
        }
        private void button1_Click(object sender, EventArgs e)
        {
            Pen mypen = ____①____ ;
            Point point1 = new Point(90, 100);
            Point point2 = new Point(165,100);
            Graphics g1 = this.CreateGraphics();
            ____②____ ;
        }
        private void button2_Click(object sender, EventArgs e)
        {
            Graphics g2 = this.CreateGraphics();
            Pen mypen = ____③____ ;
            g2.DrawLine(____④____);
        }
```

            }
    }

2. 设计一个图形用户界面的绘图程序，界面如图 2-15 所示，单击按钮后绘制与输入宽度和高度相对应的图形。请完善程序。

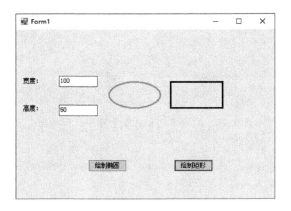

图 2-15 绘制椭圆和矩形程序运行界面

```
using System;
using System.Drawing;
using System.Windows.Forms;
namespace ex11_02
{
    public partial class Form1 : Form
    {
        public Form1()
        {
            InitializeComponent();
        }
        private void button1_Click(object sender, EventArgs e)
        {
            Graphics g1 = ____①____ ;
            Pen mypen = new Pen(Color.Cyan, 3);
            g1.DrawEllipse(mypen, 180, 100,____②____, int.Parse(textBox2.Text));
        }
        private void button2_Click(object sender, EventArgs e)
        {
            Graphics g1 = this.CreateGraphics();
            Pen mypen = ____③____ ;
            g1.DrawRectangle(mypen, 300, 100,____④____);
        }
    }
}
```

3. 设计一个图形用户界面的绘图程序，界面如图 2-16 所示，单击按钮后绘制多边形和文字图形。请完善程序。

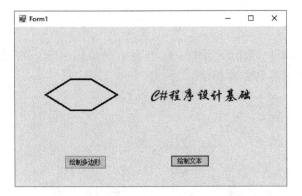

图 2-16 绘制多边形和文字程序运行界面

```
using System;
using System.Drawing;
using System.Windows.Forms;
namespace ex11_03
{
    public partial class Form1 : Form
    {
        public Form1()
        {
            InitializeComponent();
        }
        private void button1_Click(object sender, EventArgs e)
        {
            Graphics g1 = this.CreateGraphics();
            Pen mypen = new Pen(Color.Black, 3);
            Point p1 = new Point(110,100);
            Point p2 = new Point(60,130);
            Point p3 = new Point(110,160);
            Point p4 = new Point(150,160);
            Point p5 = new Point(200,130);
            Point p6 = new Point(150,100);
            Point[] mypoints =     ①     ;
            g1.DrawPolygon(mypen,     ②     );
        }
        private void button2_Click(object sender, EventArgs e)
        {
            string str = "C#程序设计基础";
            Font myfont = new Font("华文行楷", 20);
            SolidBrush mybrush = new     ③     ;
            Graphics mygraphic = this.CreateGraphics();
            mygraphic.DrawString(str, myfont,     ④     , 260, 120);
        }
    }
}
```

4. 设计一个图形用户界面的绘图程序，界面如图 2-17 所示，单击按钮后绘制图章。请完善程序。

图 2-17 绘制图章程序运行界面

```
using System;
using System.Drawing;
using System.Windows.Forms;
namespace ex11_04
{
    public partial class Form1 : Form
    {
        public Form1()
        {
            InitializeComponent();
        }
        private void button1_Click(object sender, EventArgs e)
        {
            int radius = 200;                              //定义圆的直径
            int brush_w = 3;                               //设置圆的粗细
            Rectangle rect = new Rectangle(brush_w, brush_w, radius - brush_w * 2,
                        radius - brush_w * 2);   //为绘制圆而定义的矩形
            Font star_Font = new Font("Arial", 40, FontStyle.Regular);
            Font var_font = new Font("宋体", 20, FontStyle.Bold);
            string star_str = "★";
            Graphics g1 = this.CreateGraphics();
            Pen mypen = new Pen(Color.Red, brush_w);
            g1.     ①     ;                                //在定义的rect矩形里绘制圆
            SizeF var_size = new SizeF(rect.Width, rect.Width);
            var_size = g1.MeasureString(star_str, star_Font);
            g1.DrawString(star_str, star_Font, mypen.Brush, new PointF((rect.Width /2F) +
                brush_w - var_size.Width /2F, rect.Height / 2F - var_size.Width / 2F));
            var_size = g1.MeasureString("专用章", var_font);
            g1.DrawString("专用章", var_font, mypen.Brush, new PointF((rect.Width / 2F) +
                brush_w - var_size.Width /2F, rect.Height / 2F + var_size.Height * 1.5F));
            string mystring = "云计算学术会议";
            int len = mystring.Length;
            float angle = 180 + (180 - len * 25) / 2;
            for (int i = 0; i < len; i++)                  //将文字以指定的弧度进行绘制
            {
                g1.TranslateTransform((radius + brush_w/2F)/2F, (radius + brush_w/2F)/2F);
```

```
                g1.RotateTransform(angle);              //在绘制时旋转一个角度
                Brush mybrush = Brushes.Red;
                g1.DrawString(mystring.Substring(i, 1),___②___, mybrush,60, 0);
                                                        //以 var_font 定义的字体显示旋转文字
                g1.ResetTransform();
                angle += 25;                            //设置下一个文字的角度
            }
        }
    }
}
```

5. 设计一个图形用户界面程序，界面如图 2-18 所示，选择相应菜单可以分别完成打开图像文件并显示在窗口中，将彩色图像转换成灰度图像并保存图像到文件的操作。请完善程序。

图 2-18　图像处理程序运行界面

```
using System;
using System.Drawing;
using System.Windows.Forms;
namespace ex11_05
{
    public partial class Form1 : Form
    {
        public Form1()
        {
            InitializeComponent();
        }
        private void 打开ToolStripMenuItem_Click(object sender, EventArgs e)
        {
            openFileDialog1.Filter = "All Image Files|*.bmp;*.jpg;*.gif;*.png";
            this.pictureBox1.SizeMode = PictureBoxSizeMode.AutoSize;
            if (openFileDialog1.ShowDialog() == DialogResult.OK)
            {
                this.pictureBox1.Image = Image.FromFile(___①___);
                this.Width = this.pictureBox1.Image.Width;
                this.Height = this.pictureBox1.Image.Height;
```

```csharp
        }
    }
    private void 图像转换ToolStripMenuItem_Click(object sender, EventArgs e)
    {
        Bitmap myimage = new Bitmap(pictureBox1.Image);
        Graphics g1 = pictureBox1.CreateGraphics();
        for (int i = 0; i < myimage.Width; i++)
        {
            for (int j = 0; j < myimage.Height; j++)
            {
                Color c1 = myimage.GetPixel(i, j);
                byte gray1 = (byte)((c1.R + c1.G + c1.B) /3);
                myimage.SetPixel(i, j, Color.FromArgb(gray1, gray1, gray1));
            }
        }
        g1.DrawImage(myimage, 0, 0);
        pictureBox1.Image = myimage;
        g1.Dispose();
    }
    private void 保存ToolStripMenuItem_Click(object sender, EventArgs e)
    {
        if (saveFileDialog1.ShowDialog() == _____②_____ )
            this.pictureBox1.Image.Save(saveFileDialog1.FileName);
    }
}
```

三、独立编程实验

1. 设计一个简易的 Windows 绘图板程序，界面如图 2-19 所示，能够在 Windows 窗体上绘制出简单的直线、曲线、多边形、填充图形和绘制字符。

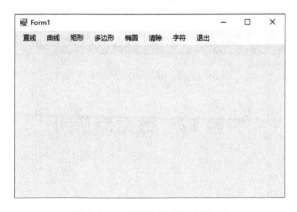

图 2-19 绘图板程序运行界面

2. 设计一个显示三原色图形的程序，程序运行界面如图 2-20 所示，单击"绘制"按钮后，绘制出表示三原色的三个圆。

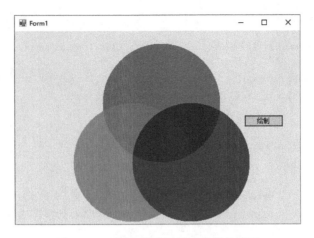

图 2-20 三原色图形显示程序运行界面

提示：定义红色填充刷。

SolidBrush redBrush = new SolidBrush(Color.FromArgb(128, 255, 0, 0));

3. 设计一个图形用户界面的程序，它具有打开图像文件、上下翻转图像、保存图像的功能，程序运行界面如图 2-21 所示。

图 2-21 图像处理程序运行界面

实验 12　数据库应用

一、实验目的

1. 掌握 ADO.NET 访问数据库的方法。
2. 掌握 ADO.NET 的数据访问对象，如 OleDbConnection、OleDbCommand 等对象的使用方法。
3. 掌握 DataSet 数据库访问组件的使用方法。
4. 掌握数据操作界面的设计方法。

二、模仿编程实验

1. 下面的程序运行界面如图 2-22 所示，单击按钮后，从 Employee 数据库的 worker(w_id, w_name, w_sex, w_birthday, w_salary, w_prize) 表中取出第一条记录显示到相应的文本框中。请完善程序。

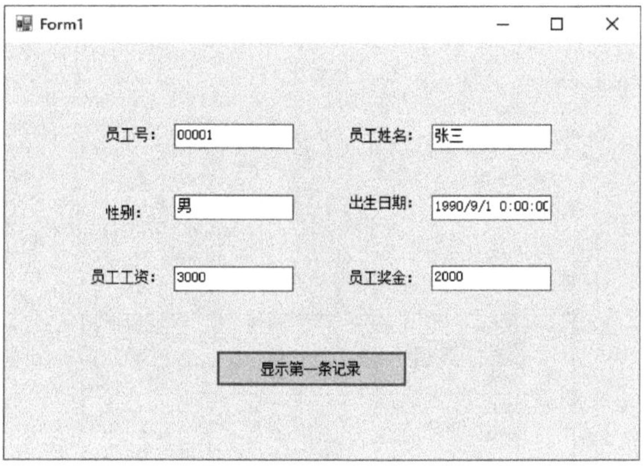

图 2-22 记录显示程序运行界面

```
using System;
using System.Data;
using System.Data.OleDb;
using System.Windows.Forms;
namespace chp12_01
{
    public partial class Form1 : Form
    {
        public Form1()
        {
            InitializeComponent();
        }
        private void button1_Click(object sender, EventArgs e)
        {
            string strconn = "Provider = SQLNCLI11;Data Source = DESKTOP - 331UI15;
                              Integrated Security = SSPI;Initial Catalog = Employee";
            OleDbConnection myconn = new OleDbConnection();
            myconn.ConnectionString = _____①_____ ;
            myconn.Open();
            OleDbDataAdapter myoledata =
                     new OleDbDataAdapter("select top 1 * from worker",myconn);
            DataTable dt = new DataTable();
            myoledata.Fill(_____②_____);
            textBox1.Text = dt.Rows[0][0].ToString().Trim();
            textBox2.Text = dt.Rows[0][1].ToString().Trim();
            textBox3.Text = dt.Rows[0][2].ToString().Trim();
            textBox4.Text = dt.Rows[0][3].ToString().Trim();
            textBox5.Text = dt.Rows[0][4].ToString().Trim();
            textBox6.Text = dt.Rows[0][5].ToString().Trim();
```

```
                myconn.Close();
            }
        }
    }
```

2. 下面的程序完成对 Employee 数据库的 worker 表的数据录入,程序界面如图 2-23 所示。请完善程序。

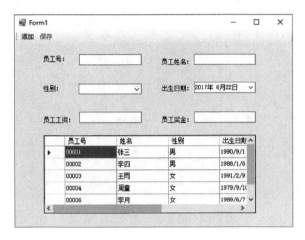

图 2-23　数据录入程序运行界面

```
using System;
using System.Data;
using System.Data.SqlClient;
using System.Windows.Forms;
namespace chp12_02
{
    public partial class Form1 : Form
    {
        string strconn = "Data Source = DESKTOP - 331UI15;
                          Initial Catalog = Employee;Integrated Security = True";
        SqlConnection myconn = new SqlConnection();
        SqlCommand mycomm = new SqlCommand();
        string strcomm;
        public Form1()
        {
            InitializeComponent();
        }
        void showinfo()
        {
            DataSet myds = new DataSet();
            SqlDataAdapter mydap = new SqlDataAdapter();
            SqlCommand mycmd = new SqlCommand();
            mycmd.Connection = myconn;
            mycmd.CommandText = "select * from worker";
            mycmd.ExecuteNonQuery();
            mydap.SelectCommand = mycmd;
            mydap.Fill(myds, "worker");
```

```
            dataGridView1.DataSource = _____①_____ ;
        }
        private void Form1_Load(object sender, EventArgs e)
        {
            myconn.ConnectionString = strconn;
            myconn.Open();
            toolStripButton1.Enabled = true;
            toolStripButton2.Enabled = false;
            showinfo();
        }
        private void toolStripButton1_Click(object sender, EventArgs e)
        {
            textBox1.Text = "";
            textBox2.Text = "";
            textBox5.Text = "";
            textBox6.Text = "";
            comboBox1.SelectedText = "男";
            dateTimePicker1.Value = DateTime.Now;
            toolStripButton2.Enabled = true;
            toolStripButton1.Enabled = false;
        }
        private void toolStripButton2_Click(object sender, EventArgs e)
        {
            if (textBox1.Text.Trim() != "")
            {
                strcomm = String.Format("insert into worker
                    values('{0}','{1}','{2}','{3}','{4}','{5}')",
                    textBox1.Text.Trim(), textBox2.Text.Trim(), comboBox1.Text.Trim(),
                    dateTimePicker1.Value.ToString(), textBox5.Text.Trim(),
                    textBox6.Text.Trim());
                mycomm.Connection = myconn;
                mycomm.CommandText = strcomm;
                _____②_____ ;
                showinfo();
                toolStripButton1.Enabled = true;
                toolStripButton2.Enabled = false;
            }
            else
                MessageBox.Show("请输入数据", "提示");
        }
    }
}
```

三、独立编程实验

1. 设计一个基于 Windows 界面的个人书籍管理系统，实现对表 Book（bookId，bookName，Author，Publisher，publishDate，Price）的添加、修改和删除功能。

2. 设计一个基于 Windows 界面的个人书籍查询系统，可以对表 Book（bookId，bookName，Author，Publisher，publishDate，Price）中的数据，按不同的条件进行查询。

第3章 常用算法设计

面向对象程序设计的核心是从需要解决的问题中抽象出合适的类,并将数据和对数据的操作方法封装在类的内部。尽管面向对象程序设计的设计思想不同于结构化程序设计,但两者并不是对立的,在面向对象程序设计中仍然要用到结构化程序设计的知识。例如,一个类的方法就要用结构化程序设计来实现。所以在学习 C♯ 程序设计时,算法设计即如何确立编写程序的思路,仍然是不能忽视的问题。算法设计是学习高级语言程序设计的难点,也是学习的重点。初学者普遍感到头疼的问题是,碰到一个问题后不知从何下手,难以建立起明确的编程思路。针对这一普遍问题,本章根据教学基本要求,将常见的程序设计问题分为累加与累乘问题、数字问题、数值计算、数组的应用和函数的应用 5 类,分别总结每一类程序设计问题的思路,以引导读者掌握基本的程序设计方法和技巧。

3.1 累加与累乘问题

累加与累乘问题是很典型、最基本的一类算法,实际应用中很多问题都可以归结为累加与累乘问题。先看累加问题。

累加的数学递推式为:

$$S_0 = 0$$
$$S_i = S_{i-1} + X_i \quad (i = 1, 2, 3, \cdots)$$

其含义是第 i 次的累加和 S 等于第 i−1 次时的累加和 S 加上第 i 次时的累加项 X。从循环的角度讲,即是本次循环的 S 值等于上一次循环时的 S 值加上本次循环的 X 值,这可用下列赋值语句来实现:

S = S + X

显然,控制上述赋值语句重复执行若干次后,S 的值即若干个数之和。

特例 1 当 X_i 恒为 1 时,即 $S_i = S_{i-1} + 1$,S 用于计数。

特例 2 当 $X_0 = 0$,且 $X_i = X_{i-1} + 1 \ (i = 1, 2, 3, \cdots, N)$ 时,S 为 $1+2+3+\cdots+N$ 之值。

再看累乘问题,其数学递推式为:

$$P_0 = 1$$
$$P_i = P_{i-1} \cdot X_i \quad (i = 1, 2, 3, \cdots)$$

其含义是第 i 次的累乘积 P 等于第 i−1 次时的累乘积 P 乘以第 i 次时的累乘项 X。从循环的角度讲,即是本次循环的 P 值等于上一次循环时的 P 值乘以本次循环的 X 值,这可用下列赋值语句来实现:

P = P * X

显然,控制上述赋值语句重复执行若干次后,P 的值即若干个数之积。

特例 1　当 $X_1=X_2=\cdots=X_{N-1}=X_N=X$ 时,P 的值为 X^N。

特例 2　当 $X_0=0$,且 $X_i=X_{i-1}+1$ ($i=1,2,3,\cdots,N$)时,P 的值为 $N!$。

递推问题常用迭代方法来处理,即赋值语句 S=S+X 或 P=P*X 循环执行若干次。相应的算法设计思路是:

(1) 写出循环体中需要重复执行的部分。这一部分要确定两个内容:一是求每次要累加或累乘的数,二是迭代关系 S=S+X 或 P=P*X。

(2) 确定终止循环的方式。一般有事先知道循环次数的计数循环和事先不知道循环次数的条件循环两种方式,依具体情况而定。计数循环可用一个变量来计数,当达到一定循环次数后即退出循环。条件循环可根据具体情况确定一个循环的条件,当循环条件不满足时即退出循环。

(3) 确定循环初始值,即第一次循环时迭代变量的值。

(4) 重新检查,以保证算法正确无误。

一般而言,这一类问题的算法流程图基本框架如图 3-1 所示。

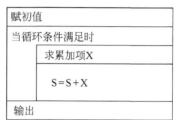

图 3-1　累加问题算法框架

【例 3-1】　已知

$$s = \sum_{i=1}^{N} \frac{2}{(4i-3)(4i-1)}$$

(1) 当 n 取 1000 时,求 s 的值。

(2) 求 s<0.78 时的最大 n 值和与此时 n 值对应的 s 值。

(3) 求 s 的值,直到累加项小于 10^{-4} 为止。

分析:第一种情况下,属于循环次数已知的循环结构,不难画出流程图如图 3-2(a)所示。第二和第三种情况下,属于循环次数未知的循环结构,根据图 3-1 的流程图框架得到流程图分别如图 3-2(b)和图 3-2(c)所示。

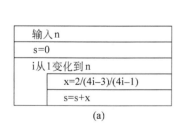
(a)

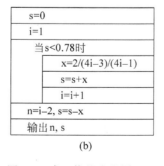
(b)　　　(c)

图 3-2　求 s 值的流程图

根据上述流程图不难分别写出程序如下:

程序 1:

```
using System;
```

```csharp
namespace exchp03_01_01
{
    class Program
    {
        static void Main(string[] args)
        {
            int i, n;
            double x, s = 0;
            Console.WriteLine("请输入 n 的值:");
            n = int.Parse(Console.ReadLine());
            for (i = 1; i < n + 1; i++)
            {
                x = 2.0 / (4 * i - 3) / (4 * i - 1);
                s += x;
            }
            Console.WriteLine("s = {0:f6}", s);
        }
    }
}
```

程序运行结果如下：

请输入 n 的值:1000 ✓
s = 0.785273

程序 2：

```csharp
using System;
namespace exchp03_01_02
{
    class Program
    {
        static void Main(string[] args)
        {
            int i = 1;
            double S = 0;
            double copyS = 0;
            while (S < 0.78)
            {
                copyS = S;
                S += 2.0 / (4.0 * i - 3.0) / (4.0 * i - 1.0);
                i++;
            }
            Console.WriteLine("When the i equals to {0}, the S is {1:f6}",
                              i - 2, copyS);
        }
    }
}
```

程序运行结果如下：

When the i equals to 23, the S is 0.779964

程序 3：
```
using System;
namespace exchp03_01_03
{
    class Program
    {
        static void Main(string[] args)
        {
            int i = 1;
            double x, s = 0;
            x = 2.0 / (4.0 * i - 3.0) / (4.0 * i - 1.0);
            for (; x >= 1e - 4;)
            {
                s += x;
                i++;
                x = 2.0 / (4.0 * i - 3.0) / (4.0 * i - 1.0);
            }
            Console.WriteLine("s = ,{0:f6}", s);
        }
    }
}
```

程序运行结果如下：

s = 0.781827

【例 3-2】 已知

$$S = \sum_{i=1}^{N} \frac{\sin(x_i + y_i)}{1 + \sqrt{x_i y_i}}$$

其中，$x_i = \begin{cases} i & (i \text{ 为奇数}) \\ \frac{i}{2} & (i \text{ 为偶数}) \end{cases}$，$y_i = \begin{cases} i^2 & (i \text{ 为奇数}) \\ i^3 & (i \text{ 为偶数}) \end{cases}$。从键盘输入 N 的值，求 S 的值。

程序如下：

```
using System;
namespace exchp03_02
{
    class Program
    {
        static void Main(string[] args)
        {
            int i = 0, N = 0;
            double xi, yi, S = 0.0;
            N = int.Parse(Console.ReadLine());
            for (i = 1; i <= N; i++)
            {
                if (i % 2 == 0)
                { xi = i / 2; yi = i * i * i; }
                else
                { xi = i; yi = i * i; }
```

```
                S += Math.Sin(xi + yi) / (1.0 + Math.Sqrt(xi * yi));
            }
            Console.WriteLine("When N = {0}, the S = {1:f6}",N,S);
        }
    }
}
```

程序运行结果如下：

30 ↙
When N = 30, the S = 0.375001

【例 3-3】 求 $y=f(1)+f(2)+\cdots+f(n)$，其中 $f(n)=(-1)^n \sqrt{2n^2+1}$。

(1) 当 n=50 时，y 的值是多少？

(2) 当 n=100 时，y 的值是多少？

程序如下：

```
using System;
namespace exchp03_03
{
    class Program
    {
        static void Main(string[] args)
        {
            int n, i;
            double y = 0.00, j = -1;
            Console.WriteLine("输入 n 的值:");
            n = int.Parse(Console.ReadLine());
            for (i = 1; i < n + 1; i++)
            {
                y += j * Math.Sqrt(2.0 * i * i + 1.0);
                j *= -1;
            }
            Console.WriteLine("y = {0:f6}", y);
        }
    }
}
```

程序运行结果如下：

输入 n 的值:50 ↙
y = 35.145424
输入 n 的值:100 ↙
y = 70.499022

【例 3-4】 已知

$$\cos x = 1 - \frac{x^2}{2!} + \frac{x^4}{4!} - \frac{x^6}{6!} + \cdots$$

输入 x 的值，计算 cosx 的值，直到最后一项的绝对值小于 10^{-5} 为止。

程序如下：

```csharp
using System;
namespace exchp03_04
{
    class Program
    {
        static double Rank(int n)
        {
            double s = 1;
            for (int i = 1; i <= n; i++)
                s *= i;
            return s;
        }
        static void Main(string[] args)
        {
            double x = 0.0, cosx = 0.0, copyx = 0.0;
            const double LIMIT = 0.00001;
            int j = 0, i = 0;
            x = double.Parse(Console.ReadLine());
            copyx = x;
            x = x * Math.PI / 180.0;
            do
            {
                if (j % 2 == 0)
                    cosx += Math.Pow(x, i) / Rank(i);
                else
                    cosx -= Math.Pow(x, i) / Rank(i);
                i += 2;
                j++;
            } while (Math.Abs(Math.Pow(x, i) / Rank(i)) >= LIMIT);
            Console.WriteLine("When x = {0}, the cosx = {1:f6}", copyx, cosx);
        }
    }
}
```

程序运行结果如下：

47✓
When x = 47, the cosx = 0.681993

3.2 数字问题

数字问题主要研究整数的一些自身性质与相互关系。处理过程中常常要用到求余数、分离数字及判断整除等技巧，务必熟练掌握。

(1) 判断一个整数 m 能否被另一个整数 n 整除。

方法 1：若 m%n 的值为 0，则 m 能被 n 整除，否则不能。

方法 2：若 m－m/n*n 的值为 0，则 m 能被 n 整除，否则不能。

(2)分离自然数 m 各位的数字。

m%10 的值是 m 的个位数字,m/10%10 的值是 m 的十位数字,以此类推,可以得到 m 的更高位数字。

数字问题的提法往往是,求某一范围内符合某种条件的数。这一类问题的算法设计思路是:

(1)考虑判断一个数是否满足条件的算法,有时候可以直接用一个关系表达式或逻辑表达式来判断,如判断奇数、偶数。但更多的情况是无法直接用一个条件表达式来判断,这时可根据定义利用一个循环结构来进行判断,例如判断一个数是否素数。

(2)在指定范围内重复执行"判断一个数是否满足条件"的程序段,从而求得指定范围内全部符合条件的数。这里用的方法是穷举。

一般而言,这一类问题的算法流程图基本框架如图 3-3 所示。

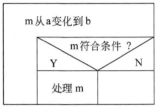

图 3-3　数字问题算法框架

【例 3-5】 考察[1000,2000]范围内的全部素数。

(1)最小的素数。

(2)由小到大第 50 个素数。

(3)全部素数之和。

程序如下:

```
using System;
namespace exchp03_05
{
    class Program
    {
        static void Main(string[] args)
        {
            int i, j, count = 0; long s = 0;
            for (i = 1000; i < 2001; i++)
            {
                for (j = 2; j < i; j++)
                    if (i % j == 0) break;
                if (j == i)
                {
                    count++;
                    if (count == 1)
                        Console.WriteLine("最小素数是,{0}",i);
                    if (count == 50)
                        Console.WriteLine("第 50 个素数是{0}",i);
                    s += i;
                }
            }
            Console.WriteLine("素数和是{0}",s);
        }
    }
}
```

程序运行结果如下：

最小素数是 1009
第 50 个素数是 1361
素数和是 200923

【例 3-6】 若两个素数之差是 2，则称这两个素数是一对孪生数。例如，3 和 5 是一对孪生数。求[2,500]中孪生数的对数和最大的一对孪生数。

程序如下：

```
using System;
namespace exchp03_06
{
    class Program
    {
        static int IsPrime(int n)
        {
            int i;
            for (i = 2; i < n; i++)
                if (n % i == 0) return 0;
            return 1;
        }
        static void Main(string[] args)
        {
            int n, count = 0, twin = 0;
            for (n = 3; n <= 497; n++, n++)
                if (IsPrime(n) == 1 && IsPrime(n + 2) == 1)
                {
                    count++;
                    twin = n;
                }
            if (twin == 0)
                Console.WriteLine("No twins in [2 500]");
            else
            {
                Console.WriteLine("The numeber of twins prime is {0}",
                                count);
                Console.WriteLine("The bigest numeber of twins prime
                                in [2,500] is {0} and {1}", twin, twin + 2);
            }
        }
    }
}
```

程序运行结果如下：

The numeber of twins prime is 24
The bigest numeber of twins prime in [2,500] is 461 and 463

【例 3-7】 若正整数 N 的所有因子之和等于 N 的倍数,则称 N 为红玫瑰数。如 28 的因子之和为 1+2+4+7+14+28=56=28×2,故 28 是红玫瑰数。求 [1,700] 中最大的红玫瑰数,[1,700] 中有多少个红玫瑰数。

程序如下:

```
using System;
namespace exchp03_07
{
    class Program
    {
        static void Main(string[] args)
        {
            int i, j, sum, count = 0, maxRose = 0;
            for (i = 1; i <= 700; i++)
            {
                sum = 0;
                for (j = 1; j <= i; j++)
                    if (i % j == 0) sum += j;
                if (sum % i == 0)
                {
                    maxRose = i;
                    count++;
                }
            }
            Console.WriteLine("The maximal red rose number is {0}",
                              maxRose);
            Console.WriteLine("There are {0} groups red rose numbers",
                              count);
        }
    }
}
```

程序运行结果如下:

```
The maximal red rose number is 672
There are 6 groups red rose numbers
```

【例 3-8】 求 [2,1000] 中因子(包括 1 和该数本身)个数最多的数及其因子个数。

程序如下:

```
using System;
namespace exchp03_08
{
    class Program
    {
        static void Main(string[] args)
        {
            int i, j, count, num = 0, maxFactor = 0;
            for (i = 2; i <= 1000; i++)
            {
                count = 0;
```

```
            for (j = 1; j <= i; j++)
                if ((i % j) == 0)
                    count++;
            if (count > maxFactor)
            {
                maxFactor = count;
                num = i;
            }
        }
        Console.WriteLine("The number {0} has
                    the maximal factors {1}",num,maxFactor);
    }
}
```

程序运行结果如下:

The number 840 has the maximal factors 32

3.3 数值计算问题

数值计算是"计算方法"课程研究的对象,主要研究如何用计算机来求一些数学问题的数值解。目前数值计算方法已趋于完臻和成熟,许多问题都有了现成的算法或软件包。详细内容可参阅数值分析或计算方法方面的文献或直接使用有关软件,如 MATLAB 科学计算软件。

【例 3-9】 用牛顿迭代法求方程 $f(x)=0$ 在 $x=x_0$ 附近的实根,直到 $x_n-x_{n-1} \leqslant \varepsilon$ 为止。

牛顿迭代公式为

$$x_n = x_{n-1} - \frac{f(x_n)}{f'(x_{n-1})}$$

本质上讲,这属于递推问题,采用迭代方法不难得到如图 3-4 所示的算法。

设 $f(x)=x^2-a$,则迭代公式为

$$x_n = \frac{1}{2}\left(x_{n-1} + \frac{a}{x_{n-1}}\right)$$

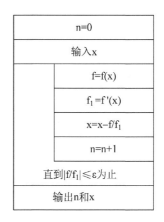

图 3-4 牛顿迭代法求方程的根

显然此时方程的根即 $\sqrt{a}$,利用此迭代公式可以求 $\sqrt{a}$ 的近似值。假定取 $x_0=a/2, \varepsilon=10^{-4}$,程序如下:

```
using System;
namespace exchp03_09
{
    class Program
    {
        static void Main(string[] args)
        {
            int n = 0;
```

```
            double a, x, x1;
            Console.WriteLine("输入a的值:");
            a = double.Parse(Console.ReadLine());
            x = a / 2.0;
            x1 = (x + a / x) / 2.0;
            while (Math.Abs(x1 - x) >= 1e-4)
            {
                n = n + 1;
                x = x1;
                x1 = (x + a / x) / 2.0;
            }
            Console.WriteLine("x = {0:f6}", x1);
        }
    }
}
```

程序运行结果如下：

输入a的值：
2
x = 1.414214

【例 3-10】 求 $S = \int_a^b f(x)dx$ 之值。

算法1：矩形法。根据定积分的几何意义，将积分区间[a,b]n等分，n个小的曲边梯形面积之和即定积分的近似值。矩形法用小矩形代替小曲边梯形，求出各小矩形面积，然后累加之，所以本质上讲这是一个累加问题，算法如图3-5所示。

也可以先找出求几个小矩形面积之和的公式，然后根据公式编写程序。

$$S = S_1 + S_2 + \cdots + S_n$$
$$= hf(a) + hf(a+h) + \cdots + hf[a+(n-1)h]$$
$$= h\sum_{i=1}^{n} f[a+(i-1)h]$$

其中，$h = \dfrac{b-a}{n}$。

显然这是一个累加问题，不难设计算法。当 $f(x) = \sqrt{1-x^2}$ 时，程序如下：

```
using System;
namespace exchp03_10_1
{
    class Program
    {
        static double f(double x)
        {
            return Math.Sqrt(1 - x * x);
        }
        static void Main(string[] args)
```

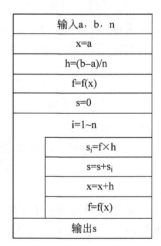

图 3-5 矩形法求定积分

```
        {
            int i;
            double sum = 0.0, a, b, h;
            int n;
            a = double.Parse(Console.ReadLine());
            b = double.Parse(Console.ReadLine());
            n = int.Parse(Console.ReadLine());
            h = (b - a) / n;
            for (i = 1; i <= 100; i++)
                sum += h * f(a + (i - 1) * h);
            Console.WriteLine("The integral is {0:f6}", sum);
        }
    }
}
```

程序运行结果如下：

-1
1
100↙
The integral is 1.569134

算法 2：梯形法。梯形法用小梯形代替小曲边梯形。

第一个小梯形的面积为

$$S_1 = \frac{f(a+h)+f(a)}{2} \cdot h$$

第二个小梯形的面积为

$$S_2 = \frac{f(a+2h)+f(a+h)}{2} \cdot h$$

…

第 i 个小梯形的面积为

$$S_i = \frac{f(a+ih)+f[a+(i-1)h]}{2} \cdot h$$

…

第 n 个小梯形的面积为

$$S_n = \frac{f(a+(n-1)h)+f(b)]}{2} \cdot h$$

本质上讲这也是一个累加问题。

也可以先找出求几个小梯形面积代数和的公式，然后根据此式设计算法。小梯形面积代数和的公式为

$$S = S_1 + S_2 + \cdots + S_n = h \cdot \frac{f(a)+f(b)}{2} + h \cdot \sum_{i=1}^{n-1} f(a+ih)$$

根据上式，当 $f(x) = \sqrt{1-x^2}$ 时，程序如下：

```
using System;
namespace exchp03_10_2
{
```

```
    class Program
    {
        static double f(double x)
        {
            return Math.Sqrt(1 - x * x);
        }
        static void Main(string[] args)
        {
            int n, i;
            double a, b, h, s;
            a = double.Parse(Console.ReadLine());
            b = double.Parse(Console.ReadLine());
            n = int.Parse(Console.ReadLine());
            h = (b - a) / n;
            s = h * (f(a) + f(b)) / 2;
            for (i = 1; i < n; i++)
                s += h * f(a + i * h);
            Console.WriteLine("s = {0:f6}", s);
        }
    }
}
```

程序运行结果如下：

-1
1
100 ↙
The integral is 1.569134

3.4 数组的应用

【例 3-11】 已知

$$\begin{cases} F_0 = F_1 = 0 \\ F_2 = 1 \\ F_n = F_{n-1} - 2F_{n-2} + F_{n-3} \quad (n > 2) \end{cases}$$

在 $F_0 \sim F_{100}$ 中求：

(1) 负数的个数。

(2) 1888 是第几项(约定 F_0 是第 0 项，F_1 是第 1 项)。

利用数组，编写程序如下：

```
using System;
namespace exchp03_11
{
    class Program
    {
        static void Main(string[] args)
        {
            int i, negativeNum = 0;
```

```
            double []F = new double[101];
            F[0] = F[1] = 0;
            F[2] = 1;
            for (i = 3; i < 101; i++)
            {
                F[i] = F[i - 1] - 2 * F[i - 2] + F[i - 3];
                if (F[i] < 0)
                    negativeNum++;
                if (F[i] == 1888)
                    Console.WriteLine("1888 是第{0}项", i);
            }
            Console.WriteLine("There are {0} negative
                        numbers together.",negativeNum);
        }
    }
}
```

程序运行结果如下：

1888 是第 29 项
There are 50 negative numbers together.

【例 3-12】 有 n 个同学围成一个圆圈做游戏，从某人开始编号（编号为 1~n），并从 1 号同学开始报数，数到 t 的同学被取消游戏资格，下一个同学（第 t+1 个）又从 1 开始报数，数到 t 的同学便第二个被取消游戏资格，如此重复，直到最后一个同学被取消游戏资格，求依次被取消游戏资格的同学编号。

程序如下：

```
using System;
namespace exchp03_12
{
    class Program
    {
        static void Main(string[] args)
        {
            const int nmax = 50;
            int i, k, m, n, t;
            int[] num = new int[nmax];
            n = int.Parse(Console.ReadLine());
            t = int.Parse(Console.ReadLine());
            for (i = 0; i < n; i++)
                num[i] = i + 1;
            i = k = m = 0;
            while (m < n)
            {
                if (num[i] != 0)
                    k++;
                if (k == t)
                {
                    Console.WriteLine("{0} is left", num[i]);
                    num[i] = 0;
```

```
                    k = 0;
                    m++;
                }
                i++;
                if (i == n)
                    i = 0;
            }
        }
    }
}
```

程序运行结果如下：

```
10
3
3 is left
6 is left
9 is left
2 is left
7 is left
1 is left
8 is left
5 is left
10 is left
4 is left
```

【例 3-13】 有一篇文章，包括 5 行，每行有若干个字符，要求统计全文中大写字母 A～Z 出现的次数。

分析：这是一个分类统计问题，容易想到的算法是采用多分支选择结构来统计出不同类别数据的个数，但当类别很多时，这样做十分烦琐。较为简便的是采用数组作分类统计，首先根据不同的类别来找到分类号，然后以分类号作为数组的下标，采取按分类号对号入座的方法，从而省去条件判断。

考虑到 26 个字母在 ASCII 码表中是连续排列的，求任一字母 ch 所对应的分类号 k 的表达式可以写成：

$$k = ch 的 ASCII 码 - 字母 A 的 ASCII 码 + 1$$

显然 ch 等于 A 时，k 的值为 1，ch 等于 B 时，k 的值为 2，…，ch 等于 Z 时，k 的值为 26。用 num 数组来作分类统计，下标变量 num(1)，num(2)，…，num(26) 分别统计字母 A，B，…，Z 的个数。

程序如下：

```
using System;
namespace exchp03_13
{
    class Program
    {
        static void Main(string[] args)
        {
            char ch;
```

```
            string str;
            int[] num = new int[26];
            int i, j, k;
            for (i = 0; i < 26; i++)
                num[i] = 0;
            for (i = 1; i <= 5; i++)
            {
                str = Console.ReadLine();
                for (j = 0; j < str.Length; j++)
                {
                    ch = str[j];
                    if (ch == '\0')
                        break;
                    if (ch >= 'A' && ch <= 'Z')
                    {
                        k = ch - 'A';
                        num[k] = num[k] + 1;
                    }
                }
            }
            for (ch = 'A',i = 0; ch <= 'Z'; ch++,i++)
            {
                if (i % 5 == 0)
                    Console.WriteLine();
                Console.Write("{0}出现的次数:{1}\t", ch, num[i]);
            };
        }
    }
}
```

程序运行结果如下：

SDFLKS
ERECVSD
ABDEKJHSDFSDF
ERERBNNM
ERDSDF

A出现的次数:1 B出现的次数:2 C出现的次数:1 D出现的次数:7 E出现的次数:6
F出现的次数:4 G出现的次数:0 H出现的次数:1 I出现的次数:0 J出现的次数:1
K出现的次数:2 L出现的次数:1 M出现的次数:1 N出现的次数:2 O出现的次数:0
P出现的次数:0 Q出现的次数:0 R出现的次数:4 S出现的次数:6 T出现的次数:0
U出现的次数:0 V出现的次数:1 W出现的次数:0 X出现的次数:0 Y出现的次数:0
Z出现的次数:0

【例3-14】 采用变化的冒泡排序法将 n 个数按从大到小的顺序排列：对 n 个数，从第一个直到第 n 个，逐次比较相邻的两个数，大者放前面，小者放后面，这样得到的第 n 个数是最小的，然后对前面 n-1 个数，从第 n-1 个到第 1 个，逐次比较相邻的两个数，大者放前面，小者放后面，这样得到的第 1 个数是最大的。对余下的 n-2 个数重复上述过程，直到按从大到小的顺序排列完毕。

程序如下：

```csharp
using System;
namespace exchp03_14
{
    class Program
    {
        static void Main(string[] args)
        {
            int[] a = new int[11];
            int i, high, low, temp;
            for (i = 1; i < 11; i++)
                a[i] = int.Parse(Console.ReadLine());
            low = 1;
            high = 10;
            while (low < high)
            {
                for (i = low; i < high; i++)
                    if (a[i] < a[i + 1])
                    {
                        temp = a[i];
                        a[i] = a[i + 1];
                        a[i + 1] = temp;
                    }
                high--;
                if (low < high)
                {
                    for (i = high; i >= low + 1; i--)
                        if (a[i] > a[i - 1])
                        {
                            temp = a[i];
                            a[i] = a[i - 1];
                            a[i - 1] = temp;
                        }
                    low++;
                }
            }
            for (i = 1; i < 11; i++)
                Console.Write("{0}\t", a[i]);
        }
    }
}
```

程序运行结果如下：

12
2
34
4
6

```
1
15
9
7
3
34      15      12      9       7       6       4       3       2       1
```

3.5 静态方法的应用

【例 3-15】 已知 $y=\dfrac{f(40)}{f(30)+f(20)}$,当 $f(n)=1\times2+2\times3+3\times4+\cdots+n\times(n+1)$ 时,求 y 的值。

程序如下:

```
using System;
namespace exchp03_15
{
    class Program
    {
        static int f(int n)
        {
            int i, sum = 0;
            for (i = 1; i <= n; i++)
                sum += i * (i + 1);
            return sum;
        }
        static void Main(string[] args)
        {
            double y = 0.0;
            y = (double)f(40) / (double)(f(30) + f(20));
            Console.WriteLine("The result is {0:f6}", y);
        }
    }
}
```

程序运行结果如下:

The result is 1.766154

【例 3-16】 一个数为素数,且依次从低位去掉 1 位、2 位、…,所得的各数仍都是素数,则称该数为超级素数,如 239。试求[100,9999]中:

(1) 超级素数的个数。
(2) 所有超级素数之和。
(3) 最大的超级素数。

程序如下:

```
using System;
namespace exchp03_16
{
```

```csharp
class Program
{
    static int IsPrime(int n)
    {
        int i;
        if (n == 1)
            return 0;
        for (i = 2; i < n; i++)
            if ((n % i) == 0)
                return 0;
        return 1;
    }
    static void Main(string[] args)
    {
        int i, count = 0, sum = 0, max = 0;
        for (i = 100; i <= 9999; i++)
        {
            if (i >= 100 && i <= 999)
            {
                if (IsPrime(i) == 1 && IsPrime(i/10) == 1 &&
                        IsPrime(i/ 100) == 1)
                {
                    count++;
                    sum += i;
                    max = i;
                }
            }
            else if (IsPrime(i) == 1 && IsPrime(i/10) == 1 &&
                        IsPrime(i/100) == 1 && IsPrime(i/1000) == 1)
            {
                count++;
                sum += i;
                max = i;
            }
        }
        Console.WriteLine("The number of the super primes is {0}", count);
        Console.WriteLine("The summation of the super primes is {0}", sum);
        Console.WriteLine("The bigest super prime is {0}", max);
    }
}
```

程序运行结果如下：

```
The number of the super primes is 30
The summation of the super primes is 75548
The bigest super prime is 7393
```

【例 3-17】 寻求并输出 3000 以内的亲密数对。亲密数对的定义为：若正整数 A 的所有因子（不包括 A）之和为 B，B 的所有因子（不包括 B）之和为 A，且 A≠B，则称 A 与 B 为亲密数对。

程序如下：

```csharp
using System;
namespace exchp03_17
{
    class Program
    {
        static int FactorSum(int n)
        {
            int i, sum = 0;
            for (i = 1; i < n; i++)
                if ((n % i) == 0)
                    sum += i;
            return sum;
        }
        static void Main(string[] args)
        {
            int i = 2;
            for (i = 2; i <= 3000; i++)
                if (i == FactorSum(FactorSum(i)) && i != FactorSum(i))
                    Console.WriteLine("{0} and {1} is the close 
                                  number group.", i, FactorSum(i));
        }
    }
}
```

程序运行结果如下：

```
220 and 284 is the close number group.
284 and 220 is the close number group.
1184 and 1210 is the close number group.
1210 and 1184 is the close number group.
2620 and 2924 is the close number group.
2924 and 2620 is the close number group.
```

3.6 解不定方程

方程的个数少于未知数的个数称为方程，这类方程没有唯一解，而有多组解。对于这类问题无法用解析法解，只能将所有可能的解一个个地去试，看是否满足方程，如满足就是方程的解，适用的方法是穷举法。

【例 3-18】 求下列不定方程组的全部正整数解。

$$\begin{cases} x+y+z=20 \\ 25x+20y+13z=400 \end{cases}$$

程序如下：

```csharp
using System;
namespace exchp03_18
{
```

```
    class Program
    {
        static void Main(string[ ] args)
        {
            int x, y, z = 0;
            for (x = 1; x <= 18; x++)
            {
                for (y = 1; y <= 18; y++)
                {
                    z = 20 - x - y;
                    if ((25 * x + 20 * y + 13 * z) == 400)
                    {
                        Console.WriteLine("The result is");
                        Console.WriteLine("x = {0},y = {1},z = {2}", x, y, z);
                    }
                }
            }
        }
    }
}
```

程序运行结果如下：

```
The result is
x = 7,y = 8,z = 5
```

【例 3-19】 求满足 A·B=716 699，且 A＋B 最小的 A 和 B。

程序如下：

```
using System;
namespace exchp03_19
{
    class Program
    {
        static void Main(string[ ] args)
        {
            int a, b, s = 716699, A = 0, B = 0;
            for (a = 1; a <= Math.Sqrt((double)716699); a += 2)
            {
                if (716699 % a == 0)
                {
                    b = 716699 / a;
                    if (s > a + b)
                    {
                        s = a + b;
                        A = a;
                        B = b;
                    }
                }
            }
            Console.WriteLine("A,B 分别为:{0},{1}", A, B);
        }
```

        }
    }

程序运行结果如下:

A,B 分别为:563,1273

思考题及答案

思考题

1. 从键盘输入 N 的值,求 $Z = \sum_{i=1}^{N}(X_i - Y_i)^2$。其中:

$$X_i = \begin{cases} i & (i \text{ 为奇数}) \\ \dfrac{i}{2} & (i \text{ 为偶数}) \end{cases}, \quad Y_i = \begin{cases} i^2 & (i \text{ 为奇数}) \\ i^3 & (i \text{ 为偶数}) \end{cases}$$

(1) 当 N 取 10 时,求 Z 的值。
(2) 当 N 取 15 时,求 Z 的值。

2. 已知

$$y = 1 + \frac{1}{2} + \frac{1}{4} + \cdots + \frac{1}{2n}$$

(1) 求 y>4 时的最小 n 值。
(2) 求与(1)的 n 值对应的 y 值。

3. 已知

$$y = \frac{e^{0.3x} - e^{-0.3x}}{2} \cdot \sin(x + 0.3)$$

(1) 当 x 取 -2.00 时,求 y 的值。
(2) 当 x 取 -3.0, -2.9, -2.8, ···, 2.9, 3.0 时,求各点 y 值之和。

4. 已知

$$e^x = 1 + x + \frac{x^2}{2!} + \frac{x^3}{3!} + \cdots + \frac{x^n}{N!}$$

若 x 取 0.5,N 取 100,求 e^x 的值。

5. 求 $S_n = a + aa + aaa + \cdots + \underbrace{aa\cdots a}_{n \uparrow a}$ 的值,其中 a 为 1~9 的整数。

提示:累加项的递推关系为 $x_n = 10x_{n-1} + a$。

6. 若一个正整数有偶数个不同的真因子,则称该数为幸运数。如 4 有 2 个真因子 1 和 2,故 4 是幸运数。求 [2,100] 中全部幸运数之和。

7. 若两个连续自然数的乘积减 1 是素数,则称这两个连续自然数是和谐数对,该素数是和谐素数。例如,2×3-1=5,由于 5 是素数,所以 2 和 3 是和谐数对,5 是和谐素数。求 [2,50] 区间内:

(1) 和谐数对的对数。
(2) 与上述和谐数对对应的所有和谐素数之和。

8. 已知 24 的 8 个因子为 1,2,3,4,6,8,12,24,而 24 正好能被 8 整除,求[1,100]中:

(1) 有多少个整数能被其因子的个数整除。

(2) 符合(1)的最大整数。

(3) 符合(1)的所有整数之和。

9. 梅森尼数是指使得 2^n-1 为素数的数 n,求[1,21]内:

(1) 有多少个梅森尼数。

(2) 最大的梅森尼数。

(3) 第二大的梅森尼数。

10. 倒勾股数是满足以下公式的 3 个整数 A,B,C。

$$\frac{1}{A^2}+\frac{1}{B^2}=\frac{1}{C^2} \quad (A>B>C)$$

(1) A,B,C 之和小于 100 的倒勾股数有多少组?

(2) 在(1)中 A,B,C 之和最小的是哪一组?

11. 满足下列两个条件的四位正整数称为四位平方数。

(1) 千位数字与百位数字相同(非 0),十位数字与个位数字相同。

(2) 是某两位数的平方。

例如,由于 $7744=88^2$,所以 7744 为四位平方数。

(1) 求所有四位平方数的数目。

(2) 求所有四位平方数之和。

12. 利用下列迭代公式计算 $y=\sqrt[3]{x}$ 的值,其中 x 的值由键盘输入,初始值 $y_0=x$,误差要求 $\varepsilon=10^{-4}$。

$$y_{n+1}=\frac{2}{3}y_n+\frac{x}{3y_n^2}$$

13. 用牛顿迭代法求方程 $e^{-x}-x=0$ 在 $x=-2$ 附近的一个实根,直到满足 $|x_{n+1}-x_n|\leqslant 10^{-6}$ 为止。

(1) 求方程的根。

(2) 求当迭代初值为-2 时的迭代次数。

14. 已知 $f(t)=\sqrt{\cos t+4\sin(2t)+5}$,求 $s=\int_0^{2\pi}f(t)dt$。

(1) 将积分区间 100 等分,利用矩形法求 s。

(2) 将积分区间 100 等分,利用梯形法求 s。

15. 已知

$$g(x)=\frac{f(f(x)+1)}{f(x)+f(2x)}$$

其中,$f(t)=\begin{cases}\dfrac{t}{1+\dfrac{t}{2}}, & 1\leqslant t\leqslant 10 \\ 2t^2+3t-5, & 其他\end{cases}$

(1) 求 g(2.5)。

(2) 求 g(17.5)。

16. 一个自然数是素数,且它的数字位置经过任意对换后仍为素数,则称为绝对素数,如13。试求所有两位绝对素数。

17. 求方程 $3x-7y=1$,在 $|x|\leqslant 100$,$|y|\leqslant 50$ 内的整数解。
(1) 共有多少组整数解。
(2) 在上述各组解中,$|x|+|y|$ 最大值是多少?
(3) 在上述各组解中,$x+y$ 最大值是多少?

18. 求满足以下条件的 x,y,z。
(1) $x^2+y^2+z^2=51^2$。
(2) $x+y+z$ 之值最大。
(3) x 最小。

思考题参考答案

1. (1) 1 304 735　　　　(2) 11 841 724
2. (1) 227　　　　　　　(2) 4.002 18
3. (1) 0.631 347　　　　(2) 19.0005
4. 1.648 72
5. 略
6. 384
7. (1) 28　　　　　　　(2) 21 066
8. (1) 16　　　　　　　(2) 96　　　　　　　(3) 686
9. (1) 7　　　　　　　　(2) 19　　　　　　　(3) 17
10. (1) 2　　　　　　　(2) 20,15,12
11. (1) 1　　　　　　　(2) 7744
12. 略
13. (1) 0.567 143　　　　(2) 6
14. (1) 13.2612　　　　　(2) 13.2612
15. (1) 0.404 392　　　　(2) 272.841
16. 11,13,17,31,37,71,73,79,97
17. (1) 29　　　　　　　(2) 143　　　　　　　(3) 137
18. $x=22,y=31,z=34$

第4章　习题选解

这一章按照课程内容体系,编写了大量的习题并给出了参考答案。在使用这些题解时,应重点理解和掌握与题目相关的知识点,而不要死记答案。应在阅读教材的基础上再来做题,通过做题达到强化、巩固和提高的目的。

习题 1　C♯语言概述

一、选择题

1. 在.NET Framework 中,MSIL 指(　　)。
   A. 目标代码　　　　B. 核心代码　　　　C. 类库　　　　D. 中间语言
2. 用 C♯语言编写的代码程序(　　)。
   A. 可立即执行　　　　　　　　　　B. 是一个源程序
   C. 经过编译即可执行　　　　　　　D. 经过编译解释才能执行
3. 以下叙述中正确的是(　　)。
   A. C♯语言的源程序不必通过编译就可以直接运行
   B. C♯语言中的每条可执行语句最终都将被转换成二进制的机器指令
   C. C♯源程序经编译形成的二进制代码可以直接运行
   D. C♯语言中的方法不可以单独进行编译
4. C♯编译程序是(　　)。
   A. 将 C♯源程序编译成中间语言(MSIL)的程序
   B. 将 C♯源程序编译成目标程序的程序
   C. 将 C♯源程序编译成应用软件
   D. C♯程序的机器语言版本
5. 在 Visual Studio 中,从(　　)窗口中可以查看当前项目的类和类型的层次信息。
   A. 类视图　　　　　　　　　　　　B. 解决方案资源管理器
   C. 属性　　　　　　　　　　　　　D. 资源视图
6. .NET Framework 将(　　)定义为一组规则,所有.NET 语言都应该遵守这些规则,才能创建可以和其他语言互操作的应用程序。
   A. JIT　　　　　　　　　　　　　　B. MSIL
   C. CLR　　　　　　　　　　　　　 D. ADO.NET
7. 存放 C♯源程序文件的扩展名是(　　)。
   A. sln　　　　　B. suo　　　　　C. exe　　　　　D. cs

8. C#程序的执行过程是(   )。
   A. 从程序的第一个方法开始,到最后一个方法结束
   B. 从程序的 Main 方法开始,到最后一个方法结束
   C. 从程序的第一个方法开始,到 Main 方法结束
   D. 从程序的 Main 方法开始,到 Main 方法结束
9. 关于 C#语言的基本语法,下列说法正确的是(   )。
   A. C#语言使用 using 关键字来引用.NET 预定义的名字空间
   B. 用 C#编写的程序中,Main 函数是唯一允许的全局函数
   C. C#语言中使用的标识符不区分大小写
   D. C#中一条语句必须写在一行内
10. 以下不属于.NET 编程语言的是(   )。
    A. Java   B. C#   C. VC.NET   D. VB.NET

二、填空题

1. _____是独立于 CPU 的指令集,它可以被高效地转换为特定于某种 CPU 的代码。
2. _____方法是程序的入口点,程序控制在该方法中开始和结束。
3. .NET Framework 有两个主要组件,分别是_____和.NET 基础类库。
4. C#语言虽然很多语言要素是从 C 语言继承而来,但是和面向过程的 C 语言不同,C#是一种_____的编程语言,主要用于开发可以在.NET 平台上运行的应用程序。

习题 2 C#程序的数据描述

一、选择题

1. 以下选项中属于 C#语言的简单数据类型是(   )。
   A. 复数型   B. 向量型   C. 浮点型   D. 集合型
2. 下列数据类型,不是 C#语言合法关键字的是(   )。
   A. double   B. float   C. integer   D. bool
3. 下列变量定义中合法的是(   )。
   A. double _a=1−1e−1;   B. double b=1+5e2.5;
   C. long do=0xfdaL;   D. float 2_and=1−e−3;
4. 在 C#语言中,合法的长整型常数是(   )。
   A. 0L   B. 4 962 710
   C. 0.054 838 743   D. 2.186 9e10
5. 下列常数中不能作为 C#常量的是(   )。
   A. 0xA5   B. 2.5e−2   C. 3e2   D. 'ab'
6. 装箱是把值类型转换到(   )类型。
   A. 数组   B. 引用   C. char   D. String
7. 下列类型中,(   )不属于引用类型。
   A. String   B. int   C. Class   D. Delegate

8. 如果左操作数大于右操作数,(　　)运算符返回 false。
    A. ==   B. <   C. <=   D. 以上都是
9. 下列标识符命名正确的是(　　)。
    A. exam-1   B. Main   C. _months   D. X.25
10. 要使用变量 age 来存储人的年龄,则将其声明为(　　)类型最为适合。
    A. sbyte   B. byte   C. int   D. float
11. 以下拆箱转换语句中,正确的是(　　)。
    A. object o;   int i＝(int)o;
    B. object o＝10.5;   int i＝(int)o;
    C. object o＝10.5;   float f＝(float)o;
    D. object o＝10.5;   float f＝(float)(double)o;
12. 以下赋值语句中,正确的是(　　)。
    A. short X＝50000;   B. ushort Y＝50000;
    C. long X＝1000; int Y＝x;   D. double x＝20; decimal Y＝x;
13. 设 double 型变量 x 和 y 的取值分别为 11.5 和 5.0,那么表达式 x/y＋(int)(x/y)－(int)x/y 的值为(　　)。
    A. 2.4   B. 2.5   C. 2.1   D. 2
14. 设 bool 型变量 a 和 b 的取值分别为 true 和 false,那么表达式 a&&(a‖!b) 和 a‖(a&&b)的值分别为(　　)。
    A. true   true   B. true   false
    C. false false   D. false true
15. 设 int 型变量 x 和 y 的取值分别为 3 和 2,那么执行下面语句后 z 的值为(　　)。
    int z = (x++ % y == 0) ? ++x : (x / y == 1) ? ++y : --y;
    A. 1   B. 2   C. 3   D. 4
16. 若有代数式 $\frac{7ae}{bc}$,则不正确的 C#语言表达式是(　　)。
    A. a/b/c＊e＊7   B. 7＊a＊e/b/c
    C. 7＊a＊e/b＊c   D. a＊e/c/b＊7
17. 有如下定义语句:
    int a = 6, b = 3, c = 5, d;
    则执行语句"d = a＞b ? (a＞c ? a : c) : b);"后 d 的值为(　　)
    A. 6   B. 3   C. 5   D. 不确定
18. 与数学式 $\frac{3x^n}{2x-1}$ 对应的 C#言表达式是(　　)。
    A. 3＊x^n/(2＊x－1)   B. 3＊x＊＊n/(2＊x－1)
    C. 3＊Math.Pow(x,n)＊(1/(2＊x－1)) D. 3＊Math.Pow(x,n)/2＊x－1
19. 若有代数式 $\sqrt{|y^x+\lg y|}$,则正确的 C#语言表达式是(　　)。
    A. Math.Sqrt(Math.Abs(Math.Pow(y,x)＋Math.Log10(y)))

B. Math.Sqrt(Math.Abs(Math.Pow(x,y)+Math.Log10(y)))

C. Math.Sqrt(Math.Abs(Math.Pow(x,y)+Math.Log(y)))

D. Math.Sqrt(Math.Abs(Math.Pow(y,x)+Math.Log(y)))

20. 在C#语言中,要求运算数必须是byte型的运算符是(　　)。

  A. /   B. ++   C. %   D. <<

21. 若有定义"int a=7;float x=2.5,y=4.7;",则表达式 x+a%3 * (int)(x+y)%2/4 的值是(　　)。

  A. 2.500 000 B. 2.750 000 C. 3.500 000 D. 0.000 000

22. sizeof(double)是(　　)。

  A. 一个双精度型表达式   B. 一个整型表达式

  C. 一种函数调用   D. 一个不合法的表达式

23. 若有以下定义和语句

```
char c1 = 'a', c2 = 'f';
Console.WriteLine("{0},{1}", (c2 - c1) , Convert.ToChar(c2 - 'a' + 'B'));
```

则输出结果是(　　)。

  A. 2,M   B. 5,!   C. 2,E   D. 5,G

24. 以下能正确地定义整型变量a,b和c并为其赋初值5的语句是(　　)。

  A. int a=b=c=5;   B. int a,b,c=5;

  C. int a=5,b=5,c=5;   D. a=b=c=5;

25. 下列关于单目运算符++,——的叙述中正确的是(　　)。

  A. 它们的运算对象可以是任何变量和常量

  B. 它们的运算对象可以是char型变量和int型变量,但不能是double型变量

  C. 它们的运算对象可以是int型变量,但不能是char型变量

  D. 它们的运算对象可以是char型变量、int型变量和double型变量

26. 以下不正确的叙述是(　　)。

  A. 在C#程序中,赋值运算符的优先级最低

  B. 在C#程序中,TOTAL 和 Total 是两个不同的变量

  C. 在C#程序中,%是只能用于整数运算的运算符

  D. 当用 ReadLine()方法从键盘输入数据时,若要输入数值,需要将输入的字符串转换成相应的数值

27. 设有定义"double y=0.5,z=1.5;int x=10;",则能够正确使用C#语言库函数的表达式是(　　)。

  A. Math.Exp(y)+Math.Abs(x) B. Math.Log10(y)+Math.Pow(x)

  C. Math.Sqrt(y-z)   D. Math.Sin(x,y)+Math.Exp(y-0.2)

28. C#中导入某一命名空间的关键字是(　　)。

  A. using   B. use   C. import   D. include

29. 下面代码的输出结果是(　　)。

```
int x = 5;
```

```
int y = x++;
Console.Write(y);
y = ++x;
Console.Write(y);
```

  A. 56   B. 67   C. 66   D. 57

30. 若有定义和语句：

```
int a = 5;
a++;
```

则此处表达式 a++ 的值是(　　)。

  A. 7   B. 6   C. 5   D. 4

31. 下列选项中，不属于值类型的是(　　)

  A. struct   B. Int32   C. int   D. string

32. 表达式 5＋'b'－'c'＋2*1.5－5/4L 的值为(　　)。

  A. 6.75   B. 5.75   C. 6   D. 5

33. 设有"int x=11;"，则表达式(x++*1/3)的值是(　　)。

  A. 3   B. 4   C. 3.666 667   D. 0

34. 声明 double a; int b; 下列选项中的表达式能够正确进行类型转换的是(　　)。

  A. a=(decimal)b;   B. a=b;   C. a=(int)b;   D. b=a;

35. 设 a 和 b 均为 double 型变量，且 a=5.6，b=2.5，则表达式(int)a+b/b 的值是(　　)。

  A. 6.5   B. 6   C. 5.5   D. 7

二、填空题

1. 操作符＿＿＿＿被用来说明两个条件同为真的情况。

2. ＿＿＿＿运算符将左右操作数相加的结果赋值给左操作数。

3. 在 C#中，进行注释有两种方法：使用//和使用"/* */"符号对，其中＿＿＿＿只能进行单行注释。

4. 布尔型的变量可以赋值为关键字＿＿＿＿或＿＿＿＿。

5. ++和－－运算符只能用于＿＿＿＿，不能用于常量或表达式。++和－－的结合方向是＿＿＿＿。

6. "Console.WriteLine("RP");"和"Console.Write("RP");"的区别是＿＿＿＿。

7. 在 C#程序中，程序的执行总是从＿＿＿＿方法开始的。

8. C#中浮点类型包括＿＿＿＿、＿＿＿＿和＿＿＿＿。

9. 要在控制台程序运行时输入信息，可使用 Console 类的＿＿＿＿方法。

10. 如果 int x 的初始值为 5，则执行表达式 x-=5%3 之后，x 的值为＿＿＿＿。

11. 若"float f=－123.567F;　int i=(int)f;"，则 i 的值是＿＿＿＿。

12. 表达式"4*10>=65 || 8%3!=0"的值为＿＿＿＿。

13. 设 x 和 y 均为 int 型变量，且 x=1,y=2，则表达式 1.0+x/y 的值为＿＿＿＿。

14. 假设已指定 i 为 int 型变量，f 为 float 型变量，d 为 double 型变量，e 为 long 型变量，则表达式 10+'a'+i*f-d/e 的结果为＿＿＿＿类型。

15. 数学式 $\sin^2 x \cdot \dfrac{x+y}{x-y}$ 写成 C♯ 语言表达式是_____。

16. C♯ 的字符常量是用_____括起来的_____个字符,而字符串常量是用_____括起来的_____序列。

17. 表达式"15==15&&10>4+5"的结果为_____。

18. 要使用变量 age 来存储人的年龄,则将其声明为_____类型最为适合。

19. Console 标准的输入和输出设备分别是_____和_____。

20. 设 double 型变量 x 和 y 的取值分别为 12.5 和 5.0,那么表达式 x/y+(int)(x/y)-(int)x/y 的值为_____。

21. 设 bool 型变量 a 和 b 的取值分别为 true 和 false,那么表达式 a&&(a‖!b) 和 a|(a&&b) 的值分别为_____和_____。

22. 若 x 和 n 均是 int 型变量,且 x 和 n 的初值均为 5,则执行表达式 x+=n++ 后,x 的值为_____,n 的值为_____。

23. 设 int 型变量 x 和 y 的取值分别为 3 和 2,那么执行下面语句后 z 的值为_____。

    int z = (x++ % y == 0) ? x : (x / y == 1) ?y : -y;

24. 若有定义"int m=5,y=2;",则执行表达式 y-=m*=y 后的 y 值是_____。

习题 3 程序流程控制

一、选择题

1. 为了避免嵌套的条件语句的二义性,C♯ 语言规定 else 与( )配对。
   A. 编辑时在同一列的 if
   B. 其之前最近的还没有配对过的 if
   C. 其之后最近的 if
   D. 同一行上的 if

2. 关于如下程序结构的描述中,正确的是( )。

   for(; ;)
   { i = i + 1; }

   A. 不执行循环体
   B. 一直执行循环体,即死循环
   C. 执行循环体一次
   D. 程序不符合语法要求

3. 下列关于 break 语句的描述中,不正确的是( )。
   A. break 语句可用在循环体中,它将使执行流程跳出本层循环体
   B. break 语句在一个循环体内可以出现多次
   C. break 语句可用在 switch 语句中,它将使执行流程跳出当前的 switch 语句
   D. break 语句可用在 if 语句中,它将使执行流程跳出当前的 if 语句

4. 下列关于 switch 语句的描述中,正确的是( )。
   A. switch 语句中 default 子句可以没有,也可以有一个
   B. switch 语句中每个语句序列中必须有 break 语句
   C. switch 语句中 case 子句后面的表达式只能是整型表达式
   D. switch 语句中 default 子句只能放在最后

5. 下列关于循环的描述中,错误的是( )。

　　A. do-while、while 和 for 循环中的循环体均可以由空语句组成

　　B. while 循环是先判断表达式,后执行循环体语句

　　C. do-while、while 和 for 循环均是先执行循环体语句,后判断表达式

　　D. do-while 循环体至少无条件执行一次,而 while 循环体可能一次也不执行

6. 下列关于 do-while 语句的描述中,正确的是( )。

　　A. do-while 语句所构成的循环只能用 break 语句跳出

　　B. do-while 语句所构成的循环不能用其他语句构成的循环来代替

　　C. do-while 语句所构成的循环只有在 while 后面的表达式为 true 时才结束

　　D. do-while 语句所构成的循环只有在 while 后面的表达式为 false 时才结束

7. 下列关于 for 循环的描述中,正确的是( )。

　　A. for 循环的循环体语句中,可以包含多条语句,但必须用花括号{ }括起来

　　B. 在 for 循环中可使用 continue 语句结束循环,接着执行 for 语句的后继语句

　　C. for 循环是先执行循环体语句,后判断表达式

　　D. for 循环只能用于循环次数已经确定的情况

8. 假定所有变量均已正确说明,下列程序段运行后,y 的值是( )。

```
int a, b, c;
a = b = c = 0; y = 21;
if (a == 0)
    y--;
else
    if (b != 0) ;
        if (c != 0)
            y = 2;
        else
            y = 1;
```

　　A. 21　　　　　　B. 1　　　　　　C. 20　　　　　　D. 2

9. 下列程序段的输出是( )。

```
int x = 3, y = -2, z = 5;
if (x < y)
    if (y < z)
        z = 2;
    else
        z += 3;
Console.WriteLine("{0}", z);
```

　　A. 2　　　　　　B. 3　　　　　　C. 5　　　　　　D. 8

10. 下列语句在控制台上的输出是( )。

```
if (true)
    Console.WriteLine("FirstMessage");
Console.WriteLine("SecondMessage");
```

A. 无输出 B. FirstMessage
C. SecondMessage D. FirstMessage
SecondMessage

11. 若 j 为整型变量,则以下循环的执行次数是( )。

```
for(int j = 2;j == 0;j--) Console.WriteLine("{0}", j);
```

A. 0 B. 1 C. 2 D. 无限次

12. 执行语句序列:

```
int y = 3;
do
{
    y -= 2;
    Console.WriteLine("{0}", y);
} while ( --y == 0);
```

输出结果是( )。

A. 死循环 B. 1—2 C. 30 D. 1

13. 执行语句序列:

```
int m = 0;
while (m < 28)
    m += 5;
Console.WriteLine("{0}", m);
```

输出结果是( )。

A. 27 B. 28 C. 30 D. 35

14. 如果 x=45,y=110,下面代码的输出结果是( )。

```
if (x < -20 || x > 40)
    if (y >= 100)
        Console.WriteLine("危险");
    else
        Console.WriteLine("报警");
else
    Console.WriteLine("安全");
```

A. 危险 B. 报警 C. 报警 安全 D. 危险 安全

15. 下列语句段将输出字符'#'的个数为( )。

```
int k = 100;
while (true)
{
    k--;
    if (k == 0) break;
    Console.WriteLine("#");
}
```

A. 101 B. 100 C. 99 D. 98

16. 判断字符变量 c 的值不是数字也不是字母,应采用表达式(　　)。

    A. c<='0' || c>='9' && c<='A' || c>='Z' && c<='a' || c>='z'
    B. !(c<='0' || c>='9' && c<='A' || c>='Z' && c<='a' || c>='z')
    C. c>='0' && c<='9' || c>='A' && c<='Z' || c>='a' && c<='z'
    D. !(c>='0' && c<='9' || c>='A' && c<='Z' || c>='a' && c<='z')

17. 能正确表示"当 x 的取值在[1,100]和[200,300]范围内为真,否则为假"的表达式是(　　)。

    A. （x>=1）&&（x<=100）&&（x>=200）&&（x<=300）
    B. （x>=1）||（x<=100）||（x>=200）||（x<=300）
    C. （x>=1）&&（x<=100）||（x>=200）&&（x<=300）
    D. （x>=1）||（x<=100）&&（x>=200）||（x<=300）

18. 设 x,y 和 z 是 int 型变量,且 x=3,y=4,z=5,则下面表达式中值为 false 的是(　　)。

    A. x!=0&&y!=0   B. x<=y
    C. x==0 || y+z!=0&&y-z!=0 D. !((x<y)&&z==0 || true)

19. 设 bool 型变量 a 和 b 的取值分别为 true 和 false,那么表达式 a&&(a||!b)和 a||(a&&b)的值分别为(　　)。

    A. true　true   B. true　false   C. false　false   D. false　true

20. 下面语句执行后 y 的值为(　　)。

    int x = 0, y = 0;
    while (x < 10) y += (x += 2);

    A. 10   B. 20   C. 30   D. 55

21. 以下选项中非法的表达式是(　　)。

    A. 0<=x<100   B. i=j==0
    C. （char)(65+3)   D. x+1=x+1

22. 能正确表示 a 和 b 同时为正或同时为负的逻辑表达式是(　　)。

    A. （a>=0 || b>=0)&&（a<0 || b<0)
    B. （a>=0&&b>=0)&&（a<0&&b<0)
    C. （a+b>0)&&（a+b<=0)
    D. a*b>0

23. 下面代码运行后,s 的值是(　　)。

    int s = 0;
    for (int i = 1; i < 100; i++)
    {
        if (s > 10)
            break;
        if (i % 2 == 0)
            s += i;
    }

    A. 20   B. 12   C. 10   D. 6

24. 下面语句执行后 y 的值为( )。

```
int x = 1; y = 1;
do
    y <<= (++x);
while (x<3);
```

  A. 16   B. 32   C. 64   D. 128

二、填空题

1. C#支持的循环有_____、_____、_____和_____循环。
2. 在 switch 语句中,每个语句标号所含关键字 case 后面的表达式必须是_____。
3. 在循环执行过程中,希望当某个条件满足时退出循环,使用_____语句。
4. 在 switch 语句中,_____语句是可选的,且若存在,只能有一个。
5. C#中每个 char 类型量占用_____个字节的内容。
6. 循环语句"for(int i＝30;i≥10;i=i-3)"循环次数为_____次。
7. 对于 do-while 循环结构,当 while 语句中的条件表达式的值为_____时结束循环。
8. 设 x,y,z 均为 int 型变量,描述"x 或 y 中有一个小于 z"的表达式是_____。
9. 条件"2＜x＜3 或 x＜－10"的 C#语言表达式是_____。
10. 判断 char 型变量 ch 是否为大写字母的正确表达式是_____。
11. 已知 A=7.5,B=2,C=3.6,表达式 A>B&&C>A ‖ A<B&&!(C>B)的值是_____。
12. 在 C#中,下列 for 循环的运行结果是_____。

```
for(int i = 0;i<5;i++)
{
    Console.Write(++i);
}
```

13. 在 C#中,下列代码的运行结果是_____。

```
for(int i = 6;i>0;i--)
{
    Console.Write(i--);
}
```

14. if 语句后面的表达式应该是_____表达式。
15. 条件运算符是一个_____目运算符,其结合性为_____。
16. 若有 if 语句为"if (a<b) min=a; else min=b;",可用条件运算符来处理的等价式子为_____。
17. 若 w=1,x=2,y=3,z=4,则条件表达式 w<x? w:y<z? y:z 的值是_____。
18. 设有变量定义"int a=5,c=4;",则(－－a==＋＋c)? －－a:c＋＋的值是_____,此时 c 的存储单元的值为_____。
19. C#的三种控制结构是_____、_____和_____。
20. _____语句实现的功能是使程序在满足另外一个特定条件时跳出本次循环。

21. 下列程序段的输出是_____。

```
int sum = 0,j;
for(j = 1; ;j++)
{
    if(sum > 20)   break;
    if(j % 3 == 0)   sum += j;
}
Console.WriteLine("{0},{1}",j,sum);
```

22. 当执行完下面语句序列后，m、n、t 的值分别为_____。

```
int a = 5,b,c,d,m,n,t;
b = c = d = 3;
m = n = t = 0;
for(;a > b;++b)
    m++;
while(a >++c)
    n++;
do{
    t++;
}while(a > d++);
```

23. 下列程序段的输出是_____。

```
int m = 15;
while(m >= 8)
{
    if( --m % 3 == 0)
        continue;
    Console.Write("m = {0},",m-- )
}
```

三、阅读程序题

1.
```
using System;
namespace chp04_03_01
{
    class Program
    {
        static void Main(string[] args)
        {
            int m, n;
            Console.WriteLine( "Enter m and n:");
            m = int.Parse(Console.ReadLine());
            n = int.Parse(Console.ReadLine());
            while (m != n)
            {
                while (m > n)
                    m -= n;
                while (n > m)
                    n -= m;
```

```
            }
            Console.WriteLine("m = {0}", m);
        }
    }
}
```

程序运行时,输入

65 45 ↙

2.

```
using System;
namespace chp04_03_02
{
    class Program
    {
        static void Main(string[] args)
        {
            int k = 0; char c = 'A';
            do
            {
                switch (c++)
                {
                    case 'A': k++; break;
                    case 'B': k--; break;
                    case 'C': k += 3; break;
                    case 'D': k %= 2; continue;
                    case 'E': k *= 10; break;
                    default: k /= 2; break;
                }
                k++;
            } while (c < 'G');
            Console.WriteLine("k = {0}", k);
        }
    }
}
```

3.

```
using System;
namespace chp04_03_03
{
    class Program
    {
        static void Main(string[] args)
        {
            int a, b, c = 0;
            for (a = 1; a < 6; a++)
                for (b = 6; b > 1; b--)
                {
                    if ((a + b) % 3 == 2)
                    { c += a + b; Console.WriteLine("{0},{1}", a, b); }
```

```
                if (c > 10) break;
            }
            Console.WriteLine("c = {0}",c);
        }
    }
}
```

4.
```
using System;
namespace chp04_03_04
{
    class Program
    {
        static void Main(string[] args)
        {
            int i, j, p = 0;
            const int T = 7;
            for (i = 1; i <= T; i += 2)
                for (j = 2; j <= T; j++)
                    if (i + j == T) Console.Write(" + ");
                    else if (i * j == T) Console.Write(" * ");
                    else p++;
            Console.WriteLine("p = {0}",p);
        }
    }
}
```

四、程序填空题

1. Fibonacci 数列的前两个数分别是 0 和 1，从第三个数开始，每个数等于前两个数的和。求 Fibonacci 数列的前 20 个数，要求每行输出 5 个数。

```
using System;
namespace chp04_03_05
{
    class Program
    {
        static void Main(string[] args)
        {
            int f, f1, f2, i;
            Console.WriteLine("Fibonacci 数列：\n");
            f1 = 0; f2 = 1;
            Console.Write("{0,6}{1,6}",f1,f2);
            for (i = 3; i <= 20; i++)
            {
                f = ____①____ ;
                Console.Write("{0,6}",f);
                if ( ____②____ ) Console.WriteLine();
                f1 = ____③____ ; f2 = f;
            }
        }
```

            }
    }

2. 计算 500 以内能被 11 整除的自然数之和。

```
using System;
namespace chp04_03_06
{
    class Program
    {
        static void Main(string[] args)
        {
            int n = 1, s;
            _____①_____ ;
            while (true)
            {
                if (_____②_____) break;
                if (_____③_____) s += n;
                n++;
            }
            Console.WriteLine("{0}",s);
        }
    }
}
```

3. "同构数"是指这样的数：它恰好出现在平方数的右端，如 $376 \times 376 = 141\ 376$。试找出 10 000 以内的全部同构数。

```
using System;
namespace chp04_03_07
{
    class Program
    {
        static void Main(string[] args)
        {
            int n, sqr = 0;
            for (n = 1; n < 10000; n++)
            {
                if (n < 10)
                    sqr = n * n % 10;
                else if (n < 100)
                    sqr = _____①_____ ;
                else if (n < 1000)
                    sqr = n * n % 1000;
                _____②_____
                    sqr = n * n % 10000;
                if (_____③_____)
                    Console.WriteLine("{0} * {1} = {2}", n, n, n * n);
            }
        }
    }
}
```

4. 有 20 只猴子吃掉 50 个桃子。已知公猴每只吃 5 个,母猴每只吃 4 个,小猴每只吃 2 个。求出公猴、母猴和小猴各多少只。

```
using System;
namespace chp04_03_08
{
    class Program
    {
        static void Main(string[] args)
        {
            int a, b, c;
            for (a = 1; a <= 10; a++)
                for (b = 1; b <= 13; ① )
                {
                    c = _____②_____ ;
                    if ( ____③____ )
                        Console.WriteLine("公猴={0},母猴={1},小猴={2}",a,b,c);
                }
        }
    }
}
```

5. 已知等比数列的第 1 项 a=1,公比 q=2。下列程序的功能是求满足前 n 项和小于 100 时的最大 n。请完善程序。

```
using System;
namespace chp04_03_09
{
    class Program
    {
        static void Main(string[] args)
        {
            int a, q, n, sum;
            a = 1; q = 2; n = sum = 0;
            do
            {
                sum = _____①_____ ;
                ++n;
                a = a * ② ;
            } while (③);
            --n;
            Console.WriteLine("{0}", n);
        }
    }
}
```

6. 下面的程序是为某超市收银台设计的一个简单结账程序。要求输入顾客购买的若干种货物的单价、数量及实收金额,计算并输出应付货款、实收金额和找零金额清单。请完善程序。

```
using System;
namespace chp04_03_10
{
    class Program
    {
        static void Main(string[] args)
        {
            int n;                              //n 表示数量
            float d, sum = 0, rmb1, rmb2;       //d 表示单价,rmb1 表示实收金额
                                                //rmb2 表示找零
            while (true)                        //永真循环
            {
                Console.WriteLine("请输入单价和数量: ");
                d = float.Parse(Console.ReadLine());
                n = int.Parse(Console.ReadLine());
                if (n == 0)
                    _____①_____ ;             //输入"0 0"时跳出循环
                sum = sum + _____②_____ ;
            }
            Console.WriteLine(" ------------------------ ");
            Console.WriteLine("总计: {0:f6}",sum);
            Console.WriteLine("应收: {0:f6}", sum);
            Console.WriteLine(" ------------------------ ");
            Console.WriteLine("现金: ");
            rmb1 = float.Parse(Console.ReadLine());
            rmb2 = ③;
            Console.WriteLine("找零: {0:f6}", rmb2);
        }
    }
}
```

五、编写程序题

1. 输入一个正整数,求该数的阶乘。

2. 计算 0~9 中任意 3 个不相同的数字组成的三位数,共有多少种不同的组合方式。

3. 口袋中有红、绿、蓝、白、黑五种颜色的球若干个。每次从口袋中取出 3 个不同颜色的球,问有多少种取法。

4. 用循环语句编程,显示输出图 4-1 所示的菱形图案。菱形的行数由键盘输入,行数不同,菱形的大小也不同。

提示:这是一个二维图形,每一个位置上的信息是行号、列号和字符,其中的行、列号控制显示位置,字符是要显示的内容。处理二维的问题用双层循环实现比较直观。由于图形是由"＊"号构成,需要循环重复显示"＊"。我们用外层循环控制行,用内层循环控制每一行中每一个位置(列)。外层循环比较简单,循环控制变量取值是从第一行到最后一行。内层循环要根据图形的变化分别确定输出空格和"＊"号的循环次数。

图 4-1 菱形图案

习题4 面向对象编程基础

一、选择题

1. C#语言的核心是面向对象编程,所有面向对象语言都应该至少具有的三个特性是(　　)。
   A. 封装、继承和多态　　　　　　B. 类、对象和方法
   C. 封装、继承和派生　　　　　　D. 封装、继承和接口

2. C#的构造函数分为实例构造函数和静态构造函数,实例构造函数可以对(　　)进行初始化,静态构造函数只能对静态成员进行初始化。
   A. 静态成员　　　　　　　　　　B. 静态成员和非静态成员
   C. 非静态成员　　　　　　　　　D. 动态成员

3. 在C#编程中,访问修饰符控制程序对类中成员的访问,如果不写访问修饰符,类的默认访问类型是(　　)。
   A. public　　　B. private　　　C. internal　　　D. protected

4. 在C#中创建类的实例需要使用的关键字是(　　)。
   A. this　　　　B. base　　　　C. new　　　　D. as

5. 在C#语言中,方法重载的主要方式有两种,包括(　　)和参数类型不同的重载。
   A. 参数名称不同的重载　　　　　B. 返回类型不同的重载
   C. 方法名不同的重载　　　　　　D. 参数个数不同的重载

6. 在C#中创造一个对象是,系统最先执行的是(　　)中的语句。
   A. main方法　　B. 构造函数　　C. 初始化函数　　D. 字符串函数

7. 分析一下C#片段中的属性,该属性是(　　)属性。

   ```
   private string name;
   public string Name{
   get{return name;}
   }
   ```
   A. 可读可写　　B. 只写　　　　C. 只读　　　　D. 静态

8. 在以下代码中,(　　)是类Teacher的方法。

   ```
   public class Teacher
   {
       int age = 33;
       private string name;
       public string Name
       {
           get{return name;}
           set{name = value;}
       }
       public void SaySomething{
       //…
   ```

          }
    }

    A. Name B. name C. age D. SaySomething

9. 在C♯中,下列关于属性的使用正确的是( )。

   A.
   ```
   private int num;
   public string Num
   {
       get{retuen num;}
       set{num = value;}
   }
   ```

   B.
   ```
   private int num;
   public int Num
   {
       get{retuen num;}
       set{num = value;}
   }
   ```

   C.
   ```
   private int num;
   public int Num
   {
       get{ num = value;}
       set{ retuen num;}
   }
   ```

   D.
   ```
   private int num;
   private int Num
   {
       get{retuen num;}
       set{num = value;}
   }
   ```

10. 以下关于C♯中的构造函数说法正确的是( )。

    A. 构造函数可以有参数
    B. 构造函数有返回值
    C. 一般情况下,构造函数总是private类型的
    D. 构造函数可以通过类的实例调用

11. 在C♯程序中,定义如下方法,下面选项中( )错误地实现对该方法的重载。

    public string Do(int value,string s){//省略代码}

    A. public int Do(int value,string s){//省略代码}
    B. public string Do(string s,int value){//省略代码}
    C. public void Do(string s,int value) {//省略代码}
    D. public void Do(){//省略代码}

12. 下列关于重载的说法,错误的是( )。

    A. 方法可以通过指定不同的参数个数重载
    B. 方法可以通过指定不同的参数类型重载
    C. 方法可以通过指定不同的参数传递方式重载
    D. 方法可以通过指定不同的返回值类型重载

13. 在C♯中,某程序在一个类中编写了两个同名的方法,该段代码的运行结果是( )。

```
public class Musician
{
    public void Song()
    {
        Console.WriteLine("Good");
    }
```

```
public void Song(string musicName)
{
    Console.WriteLine(musicName);
}
static void Main()
{
    Musician musician = new Musician();
    string musicName = "Better";
    musician.Song("Best");
}
}
```

  A. Good   B. Better   C. Best   D. 没有输出

14. 在 C# 类中,(　　)。

  A. 允许有多个相同名称的构造函数   B. 允许有多个不相同名称的构造函数
  C. 至少要有一个构造函数   D. 只能有一个构造函数

15. C# 中 MyClass 为一自定义类,其中有以下方法定义

`public void Hello(){…}`

使用以下语句创建了该类的对象,并使变量 obj 引用该对象:

`MyClass obj = new MyClass();`

那么,访问类 MyClass 的 Hello 方法是语句(　　)。

  A. obj.Hello();   B. obj::Hello();
  C. MyClass.Hello();   D. MyClass::Hello();

二、填空题

1. 面向对象语言都应至少具有的三个特性是封装、_____和多态。

2. 声明为_____的一个类成员,只有定义这些成员的类的方法能够访问。

3. _____提供了对对象进行初始化的方法,而且它在声明时没有任何返回值。

4. _____是用一个框架把数据和代码组合在一起,形成一个对象。

5. 在类对象被释放、销毁时,系统自动调用该类的_____来完成一些诸如内存释放等的收尾工作。

6. 类的静态成员属于_____所有,非静态成员属于类的实例所有。

7. 要给属性对应的数据成员赋值,通常要使用 set 访问器,set 访问器始终使用_____来设置属性的值。

8. get 访问器必须用_____语句来返回。

9. C# 中的属性通过_____和 set 访问器来对属性的值进行读和写。

10. 在类的定义中,类的_____描述了该类的对象的行为特征。

11. C# 的类定义中可以包含两种成员:静态成员和非静态成员。使用_____关键字修饰的是静态成员,反之为非静态成员。

12. 方法重载是指类的两个或两个以上的方法_____,但形式参数列表不同的情况。

13. 类声明后,可以创建类的实例,创建类的实例要使用_____关键字。类的实例相当于一个变量。创建类的实例的格式,如类名 对象名 = _____。

14. 创建新对象时将调用类的_____。它主要用来为对象分配存储空间,完成初始化操作。

三、阅读程序题

1.

```csharp
using System;
namespace chp04_04_01
{
    class Sample
    {
        int i;
        static int k = 0;
        public  Sample(){i++; k++;}
        public void Show()    { Console.WriteLine("i = {0}, k = {1}", i, k ); }

    };
    class Program
    {

        static void Main(string[ ] args)
        {
            Sample a = new Sample();
            a.Show();
            Sample b = new Sample();
            b.Show();
        }
    }
}
```

2.

```csharp
using System;
namespace chp04_04_02
{
    class Count
    {
        public Count() { count++; }
        public int HM() { return count; }
        ~Count()
        {
            count -- ;
            Console.WriteLine("{0}\t", count);
        }
        private static int count = 6;
    };
    class Program
    {
        static void Main(string[ ] args)
        {
            Count c1 = new Count(), c2 = new Count(), c3 = new Count(), c4 = new Count();
            Console.WriteLine("{0}\t{1}\t{2}\t{3}", c1.HM(), c2.HM(), c3.HM(), c4.HM());
```

               }
          }
     }

3.

```
using System;
namespace chp04_04_03
{
    class Program
    {
        static void Main(string[] args)
        {
            MyClass m = new MyClass();
            int a, b, c;
            a = 0; b = 20; c = 10;
            m.sort(ref a, ref b, ref c);
            Console.WriteLine("a={0},b={1},c={2}", a, b, c);
        }
    }
    class MyClass
    {
        public void sort(ref int x, ref int y, ref int t)
        {
            int tmp;
            if (x > y) { tmp = x; x = y; ; y = tmp; }
            if (x > t) { tmp = x; x = t; t = tmp; }
            if (y > t) { tmp = y; y = t; t = tmp; }
            Console.WriteLine("{0},{1},{2}", x, y, t);
        }
    }
}
```

4.

```
using System;
namespace chp04_04_04
{
    public class Student
    {
        public static int Add(int a, int b)
        {
            a++;
            b++;
            return a + b;
        }
        static void Main()
        {
            int a = 0;
            int b = 2;
            int c = Add(a, b);
            Console.Write(a);
```

```
                Console.Write(b);
                Console.Write(c);
            }
        }
    }
```

5.

```
using System;
namespace chp04_04_05
{
    class Program
    {
        static void Main(string[] args)
        {
            Class1 c1 = new Class1();
            Class1.y = 5;
            c1.output();
            Class1 c2 = new Class1();
            c2.output();
        }
        public class Class1
        {
            private static int x = 0;
            public static int y = x;
            public int z = y;
            public void output()
            {
                Console.Write(Class1.x);
                Console.Write(Class1.y);
                Console.Write(z);
            }
        }
    }
```

6.

```
using System;
namespace chp04_04_06
{
    class Program
    {
        class Class1
        {
            private string str = "Class1.str";
            private int i = 0;
            static void StringConvert(string str)
            {
                str = "Hello,World!";
            }
            static void StringConvert(Class1 c)
```

```csharp
        {
            c.str = "Hello,World!";
        }
        static void Add(int i)
        {
            i++;
        }
        static void AddWithRef(ref int i)
        {
            i++;
        }
        static void Main()
        {
            int i1 = 5;
            int i2 = 10;
            string str = "str";
            Class1 c = new Class1();
            Add(i1);
            AddWithRef(ref i2);
            Add(c.i);
            StringConvert(str);
            StringConvert(c);
            Console.WriteLine(i1);
            Console.WriteLine(i2);
            Console.WriteLine(c.i);
            Console.WriteLine(str);
            Console.WriteLine(c.str);
        }
    }
}
```

7.

```csharp
using System;
namespace chp04_04_07
{   class Person
    {
        private int pAge;
        public int age
        {
            get { return pAge; }
            set {
                if (value > 0)
                    pAge = value;
                else
                    pAge = 0;
            }
        }
        public void print()
        { Console.WriteLine(pAge); }
```

            }
        class Program
        {   static void Main(string[ ] args)
            {
                Person p = new Person();
                p.age = -8;
                p.print();
                p.age = 13;
                p.print();
            }
        }
    }

四、程序填空题

1. 以下程序运行后输出"hello everyone!"。请完善程序。

```
using System;
namespace chp04_04_08
{
    class Program
    {
        static void Main(string[ ] args)
        {
            A1   a = ____①____ ;
            Console.WriteLine(____②____);
        }
    }
    class A1
    {
        public string Display()
        {
            return "hello everyone!";
        }
    }
}
```

2. 以下程序运行后输出"Hello,World"。请完善程序。

```
using System;
namespace chp04_04_09
{
    class Program
    {
        static void Main(string[ ] args)
        {
            Person p = new Person();
            ____①____ = "Hello,World";
            Console.WriteLine(p.N);
        }
    }
    public class Person
    {
```

```
            private string p_name = "hello";
            public string N
            {
                get {    ②    }
                set {    ③    }
            }
        }
    }
```

五、编写程序题

1. 设计一个矩形类,并用其成员函数计算两个给定矩形的周长和面积。

2. 创建一个 employee 类,该类中有员工姓名、部门和职称(用 string 类型定义)。在类里定义构造函数、changname()和 display()方法,并用构造函数初始化每个成员,用 display()方法打印完整的对象数据。其中的数据成员是保护的,方法是公有的。

3. 设计一个玩具类 toy,类中包含玩具名称、单价、售出数量等数据,并且可以计算每种玩具销售的总金额,为该类建立一些必要的方法,并在主程序中建立若干个带有单价和售出数量的对象,显示每种玩具销售的总金额。

4. 设计一个时钟类,可以设置和显示时间,并且具有将时间增加 1s、1min、1hour 的方法。

5. 设计一个三角形类,三角形由三个顶点坐标(私有属性),判断是否为三角形,并求解其面积。

习题 5　面向对象高级编程

一、选择题

1. 下列关于继承的说法中,选项正确的是(　　)。
    A. 派生类可以继承多个基类的方法和属性
    B. 派生类必须通过 base 关键字调用基类的构造函数
    C. 继承最主要的优点是提高代码性能
    D. 继承是指派生类可以获取其基类特征的能力

2. 在 C#中,使用(　　)访问修饰符修饰的方法被称之为抽象方法。
    A. this   B. abstract   C. new   D. virtual

3. C#代码如下,代码的输出结果是(　　)。

```
class Program
{
    static void Main(string[] args)
    {
        Student s = new Student();
        s.sayHi();
        Console.ReadLine();
    }
}
class Person
```

```
{
    public virtual void sayHi()
    {
        Console.WriteLine("你好");
    }
}
class Student:Person
{
    public override void sayHi()
    {
        Console.WriteLine("你好,我是一名学生");
    }
}
```

  A. 学生   B. 你好,我是一名学生
  C. 你好!   D. 空
   你好,我是一名学生

4. 下列关于多态的说法中,选项正确的是( )。
  A. 重写虚方法时可以为虚方法指定别称
  B. 抽象类中不可以包含虚方法
  C. 虚方法是实现多态的唯一手段
  D. 多态性是指以相似的手段来处理各不相同的派生类

5. 以下关于密封类的说法,正确的是( )。
  A. 密封类可以用作基类   B. 密封类可以是抽象类
  C. 密封类永远不会有任何派生类   D. 密封类或密封方法可以重写或继承

6. 以下关于接口的说法,不正确的是( )。
  A. 接口不能实例化
  B. 接口中声明的所有成员隐式地为 public 和 abstract
  C. 接口默认的访问修饰符是 private
  D. 继承接口的任何非抽象类型都必须实现接口的所有成员

7. 派生类访问基类的成员,可使用( )关键字。
  A. base   B. this   C. out   D. external

8. 类的以下特性中,可以用于方便地重用已有的代码和数据的是( )。
  A. 多态   B. 封装   C. 继承   D. 抽象

9. 下列关于抽象类的说法错误的是( )。
  A. 抽象类可以实例化   B. 抽象类可以包含抽象方法
  C. 抽象类可以包含抽象属性   D. 抽象类可以引用派生类的实例

10. 以下关于继承的说法错误的是( )。
  A. .NET 框架类库中,object 类是所有类的基类
  B. 派生类不能直接访问基类的私有成员
  C. protected 修饰符既有公有成员的特点,又有私有成员的特点
  D. 基类对象不能引用派生类对象

11. 继承具有（　　），即当基类本身也是某一类的派生类时，派生类会自动继承间接类的成员。

　　A. 规律性　　　　B. 传递性　　　　C. 重复性　　　　D. 多样性

12. 下列说法中，正确的是（　　）。

　　A. 派生类对象可以强制转换为基类对象

　　B. 在任何情况下，基类对象都不能转换为派生类对象

　　C. 接口不可以实例化，也不可以引用实现该接口的类的对象

　　D. 基类对象可以访问派生类的成员

13. 下面关于虚方法说法错误的是（　　）。

　　A. 使用 virtual 关键字修饰虚方法

　　B. 虚方法必须被其子类重写

　　C. 虚方法可以有自己的方法体

　　D. 虚方法和抽象方法都可以实现多态性

14. 在 C# 中，关于接口下面说法错误的是（　　）。

　　A. 接口是一组规范和标准

　　B. 接口可以约束类的行为

　　C. 接口中只能含有未实现的方法

　　D. 接口中的方法可以指定具体实现，也可以不指定具体实现

15. 在定义类时，如果希望类的某个方法能够在派生类中进一步改进，以处理不同的派生类的需要，则应该将方法声明成（　　）。

　　A. sealed 方法　　B. public 方法　　C. virtual 方法　　D. override 方法

16. 下列关于"方法重载"的描述中，不正确的是（　　）。

　　A. 方法重载可以扩充现有类的功能

　　B. 构造函数不可以重载

　　C. 方法 ConsoleW(int _value)是方法 ConsoleW(string _value)的重载

　　D. 方法重载即"同样的方法名但传递的参数不同"

17. 以下说法正确的是（　　）。

　　A. 虚方法必须在派生类中重写，抽象方法不需要重写

　　B. 虚方法可以在派生类中重写，抽象方法必须重写

　　C. 虚方法必须在派生类中重写，抽象方法必须重写

　　D. 虚方法可以在派生类中重写，抽象方法也不需要重写

18. 下列关于接口的说法，正确的是（　　）。

　　A. 接口可以被类继承，本身也可以继承其他接口

　　B. 定义一个接口，接口名必须使用大写字母 I 开头

　　C. 接口像类一样，可以定义并实现方法

　　D. 可以继承多个接口，接口只能继承一个接口

19. 在 C# 语言中，以下关于继承的说法错误的是（　　）。

　　A. 一个子类不能同时继承多个父类

　　B. 任何类都是可以被继承的

C. 子类继承父类,也可以说父类派生了一个子类

D. Object 类是所有类的基类

20. 在 C# 中,如果类 C 继承自类 B,类 B 继承自类 A,则以下描述正确的是(　　)。

    A. C 继承了 B 中的成员,同样也继承了 A 中的成员

    B. C 只继承了 B 中的成员

    C. C 只继承了 A 中的成员

    D. C 不能继承 A 或 B 中的成员

21. 关于 base 关键字,下列使用方法错误的是(　　)。

    A. 在子类中,base 可以调用父类的构造函数

    B. 在子类中,base 关键字可以访问父类的公共属性

    C. 在子类中,base 关键字不可以调用父类的 protected 成员

    D. 在子类中,base 关键字不可以调用父类的 private 成员

22.
```
public abstract class Animal
{
public abstract void Eat();
public void Sleep(){ }
}
```
以下关于 C# 代码描述正确的是(　　)。

    A. 该段代码正确

    B. 代码错误,因为类中存在非抽象方法

    C. 代码错误,因为类中方法没有实现

    D. 通过代码 Animal an = new Animal;可以创建一个 Animal 对象

23. 下列类的定义中,(　　)是合法的抽象类。

    A. sealed abstract class c1{abstract public void test() {}

    B. abstract sealed public viod test();

    C. abstract class c1 {abstract void test();

    D. abstract class c1 {abstract public void test(); }

24. 下列关于继承机制的描述中不正确的是(　　)。

    A. 提供继承机制有利于提高软件模块的可重用性及可扩充性

    B. 继承机制使面向对象的开发语言能够更准确地描述客观世界,使软件开发方式变简单

    C. 继承机制使得软件开发过程效率更高

    D. 继承机制使得软件开发的难度相对增加

25. 下列关于继承的理解,错误的是(　　)。

    A. 子类可以从父类中继承其所有的成员

    B. 无论是否声明,子类都继承自 object(System.object)类

    C. 假如,类 M 继承自类 N,而类 N 又继承自类 P,则类 M 也继承自类 P

    D. 子类应是对基类的扩展。子类可以添加新的成员,但不能除去已经继承的成员的定义

二、填空题

1. 类的 protected 类型成员只允许在_____及其子类被直接访问。
2. _____是指同一个消息或操作作用于不同的对象,可以有不同的解释,产生不同的执行结果。
3. 在声明类时,在类名前加_____修饰符,则声明的类只能作为其他类的基类,不能被实例化。
4. 继承具有_____,即当基类本身也是某一类的派生类时,派生类会自动继承间接基类的成员。
5. C#中的派生类可以从它的基类中继承_____、_____、_____、事件、索引器等。
6. 如果类是从一个基类派生而来,则在调用派生类的默认构造函数前会自动调用基类的_____。
7. 一个类不可以多重继承几个父类,但是通过_____可以实现多重继承。
8. 委托声明的关键字是_____。
9. 在定义类时,如果希望类的某个方法能够在派生类中进一步进行改进,以处理不同的派生类的需要,则应将该方法声明为_____。
10. 接口只包含方法的声明,方法的实现是在_____中完成的。

三、阅读程序题

1.

```
using System;
namespace chp04_05_01
{
    class Program
    {
        static void Main(string[ ] args)
        {
            Mouse m = new Mouse();
            m.Eat();
            m.Sleep();
        }
    }
    public abstract class Animal
    {
        public abstract void Sleep();
        public virtual void Eat()
        { Console.Write("eat something"); }
    }
    public class Mouse : Animal
    {
        public override void Sleep()
        { Console.Write("mouse sleeping!"); }
        public override void Eat()
        { Console.Write("eat cheese!"); }
```

        }
    }

2.

```
using System;
namespace chp04_05_02
{
    class Program
    {
        public class TestB
        {
            public TestB()
            {
                Console.Write("begin create B object\n");
            }
            ~TestB()
            {
                Console.Write("begin destory B object\n");
            }
        }
        public class TestA : TestB
        {
            public TestA()
            {
                Console.Write("begin create A object\n");
            }
            ~TestA()
            {
                Console.Write("begin destory A object\n");
            }
        }
        static void Main(string[] args)
        {
            TestA a = new TestA();
        }
    }
}
```

3.

```
using System;
namespace chp04_05_03
{
    class Program
    {
        static void Main(string[] args)
        {
            Elephant e1 = new Elephant("abc");
            Elephant e2 = new Elephant();
        }
    }
```

```csharp
        public class Animal
        {
            public Animal()
            {
                Console.WriteLine("基类默认构造函数!");
            }
            public Animal(string s)
            {
                Console.WriteLine("非默认构造函数");
            }
        }
        public class Elephant : Animal
        {
            public Elephant()
            {
                Console.WriteLine("派生类构造函数!");
            }
            public Elephant(string str) : base(str)
            {
                Console.WriteLine(str);
            }
        }
    }
```

4.
```csharp
using System;
namespace chp04_05_04
{
    class Program
    {
        static void Main(string[] args)
        {
            s s1 = new s();
            s t1 = new s();
        }
    }
    public class s
    {
        public s()
        { Console.WriteLine("构造函数!"); }
        static s()
        { Console.WriteLine("静态构造函数!"); }
    }
}
```

5.
```csharp
using System;
namespace chp04_05_05
{
    class Program
```

```csharp
{
    public delegate void SubEventHandler();
    public abstract class Subject
    {
        public event SubEventHandler SubEvent;
        protected void FireAway()
        {
            if (this.SubEvent != null)
                this.SubEvent();
        }
    }
    public class Cat : Subject
    {
        public void Cry()
        {
            Console.WriteLine("cat cryed.");
            this.FireAway();
        }
    }
    public abstract class Observer
    {
        public Observer(Subject sub)
        {
            sub.SubEvent += new SubEventHandler(Response);
        }
        public abstract void Response();
    }
    public class Mouse : Observer
    {
        private string name;
        public Mouse(string name, Subject sub) : base(sub)
        {
            this.name = name;
        }
        public override void Response()
        {
            Console.WriteLine(name + " attempt to escape!");
        }
    }
    public class Master : Observer
    {
        public Master(Subject sub) : base(sub) { }
        public override void Response()
        {
            Console.WriteLine("host waken");
        }
    }
    class Class1
    {
        static void Main(string[] args)
        {
```

```
            Cat cat = new Cat();
            Mouse mouse1 = new Mouse("mouse1", cat);
            Mouse mouse2 = new Mouse("mouse2", cat);
            Master master = new Master(cat);
            cat.Cry();
        }
    }
}
```

四、程序填空题

1. 下面的程序定义了一个 Shape 基类，从 Shape 基类派生出直角三角形类 Triangle 类，在 Shape 类定义了虚方法 area()，在派生类 Shape 中通过 area()方法求两条直角边分别为 3 和 4 的三角形面积。请完善程序。

```
using System;
namespace chp04_05_06
{
    class Program
    {
        static void Main(string[] args)
        {
            Triangle t = new _____①_____ ;
            double s = _____②_____ ;
            Console.WriteLine("area is {0}", s);
            Console.ReadLine();
        }
    }
    class Shape
    {
        protected double width;
        protected double height;
        public Shape()
        {
            width = height = 0;
        }
        public Shape(double w, double h)
        {
            width = w;
            height = h;
        }
        public virtual double area()
        {
            return width * height;
        }
    }
    class Triangle : _____③_____
    {
        public Triangle(double x, double y): base(x, y)
        { }
```

```
            public override double area()
            {
                return width * height / 2;
            }
        }
    }
```

2. 已知下面程序输出 A1　B2　A2　B3　C3，请填空。

```
using System;
namespace chp04_05_07
{
    class A
    {
        protected int a;
        public A(int x){
            a = x;
            Console.Write("A{0} ",a); }
    };
    class B
    {
        private int b;
        public B(int x)
        {
            ____①____;
            b = x + 1;
            Console.Write("B{0} ", b);
        }
    };
    class C : B
    {
        int c;
        public C(int x):____②____
        {
            B a2 = new B(x + 1);
            ____③____;
            Console.Write("C{0} ",c); }
    };
    class Program
    {
        static void Main(string[] args)
        {
            C doobj = new C(1);
            Console.WriteLine();
        }
    }
}
```

五、编写程序题

1. 建立一个基类 Building，用来存储一座楼房的层数、房间数以及它的总平方米数。建立派生类 Housing，继承 Building，并存储卧室和浴室的数量。另外，建立派生类 Office，继承 Building，并存储灭火器和电话的数目。编写应用程序，建立住宅楼对象和办公楼对象，并输出它们的有关数据。

2. 声明一个 Shape 基类，在此基础上派生出 Square 和 Circle 类，二者都用 SetArea()函数计算对象的面积，再使用 Square 类创建一个派生类 Rectangle。

3. 定义一个基类，含有成员姓名、性别、年龄，再由基类派生出教师类和学生类，教师类增加工号、职称和工资，学生类增加学号、班级、专业和入学成绩。

习题 6　复杂数据表示与应用

一、选择题

1. C#数组主要有三种形式，它们是（　　）。
　　A. 一维数组、二维数组、三维数组　　B. 整型数组、浮点型数组、字符型数组
　　C. 一维数组、多维数组、交错数组　　D. 一维数组、二维数组、多维数组

2. 数组 pins 的定义如下：

　　int[ ] pins = new int[4]{1,2,3,4};

则 pins[1]=（　　）
　　A. 1　　　　　　　B. 2　　　　　　　C. 3　　　　　　　D. 4

3. 在 C#中，表示一个字符串的变量应使用以下（　　）语句定义。
　　A. CString str;　　　　　　　　　　B. string str;
　　C. Dim str as string　　　　　　　　D. char * str;

4. 有说明语句"double[,] tab=new double[2,3];"，那么下面叙述正确的是（　　）。
　　A. tab 是一个数组维数不确定的数组，使用时可以任意调整
　　B. tab 是一个有两个元素的一维数组，它的元素初始值分别是 2,3
　　C. tab 是一个二维数组，它的元素个数一共有 6 个
　　D. tab 是一个不规则数组，数组元素的个数可以变化

5. 下列关于数组的描述中，（　　）选项是不正确的。
　　A. String 类中的许多方法都能用在数组中
　　B. System.Array 类是所有数组的基类
　　C. String 类本身可以被看为一个 System.Char 对象的数组
　　D. 数组可以用来处理数据类型不同的批量数据

6. 枚举类型是一组命名的常量集合，所有整型都可以作为枚举类型的基本类型，如果类型省略，则定义为（　　）。
　　A. uint　　　　　　B. sbyte　　　　　　C. ulong　　　　　　D. int

7. 在数组中对于 for 和 foreach 语句，下列（　　）选项中的说法不正确。
　　A. foreach 语句能不用索引就可以遍历整个数组
　　B. foreach 语句总是从索引 1 遍历到索引 Length
　　C. foreach 总是遍历整个数组
　　D. 如果需要修改数组元素就必须使用 for 语句

8. 在 C# 中,可以避免使用枚举变量来避免不合理的赋值,以下枚举定义中正确的是(   )。

    A. public enum Sex｛male,femal｝   B. public enum Sex｛male,femal;｝

    C. public Sex enum ｛male,femal;｝   D. public Sex enum ｛male,femal｝

9. 以下关于结构的说法,正确的是(   )。

    A. 结构不可以通过 ref 或 out 形参以引用方式传递给函数成员

    B. 结构是值类型,类是引用类型

    C. 结构和类一样,均支持继承

    D. 结构允许声明无形参的实例构造函数

10. 假定一个 10 行 20 列的二维整型数组,下列哪个定义语句是正确的(   )。

    A. int[]arr = new int[10,20]   B. int[]arr = int new[10,20]

    C. int[,]arr = new int[10,20]   D. int[,]arr = new int[20;10]

11. 以下程序的输出结果是(   )。

```
string str = "b856ef10";
string result = "";
for(int i = 0;str[i]>= 'a'&&str[i]<= 'z';i += 3)
result = str[i] + result;
Console.WriteLine(result);
```

    A. 10fe658b   B. feb   C. 10658   D. b

12. 下列语句创建了(   )个 string 对象

```
string [,]  strArray = new string[3,4]
```

    A. 0   B. 3   C. 4   D. 12

13. 在 C# 中,下列代码的运行结果是(   )。

```
using System;
class Test
{
    static void Main(string[] args)
    {
        string[] strings = {"a","b","c"};
        foreach(string info in strings)
            Console.Write(info);
    }
}
```

    A. abc   B. a   C. b   D. c

14. 在 C# 中,下列代码的运行结果是(   )。

```
int[]age = new int[]{16,18,20,14,22};
foreach(int i in age)
{
    if(i>18)
        continue;
    Console.Write(i.ToString() +",");
}
```

A. 16,18,20,14,22   B. 16,18,14,22
C. 16,18,14   D. 16,18

15. 下列关于.NET中枚举型的描述正确的是（   ）。

　　A. 枚举型是引用类型

　　B. 枚举型是值类型，它是一组称为枚举数列表的命名常量组成的独特类型

　　C. 指定分配给每个枚举数的存储大小的基础类型是可以为 int，也可以为 char 类型，还可以为 long 类型

　　D. 枚举型的默认基础类型为 int，第一个枚举数的值必须为 0，后面每个枚举型的值依次递增 1

16. 在 C# 中，下列代码的运行结果是（   ）。

```
int []age1 = new int[]{10,20};
int []age2 = age1;
age2[1] = 30;
Console.WriteLine(age1[1]);
```

A. 0   B. 10   C. 30   D. 20

二、填空题

1. System.Array 有一个_____属性，通过它可以获取数组的长度。

2. 在 C# 语言中，可以用来遍历数组元素的循环语句是_____。

3. 数组是一种_____类型。

4. _____是数组的数组，它内部每个数组的长度可以不同，就像一个锯齿形状。

5. 每个枚举成员均具有相关联的常量值，默认第一个枚举成员的关联值为_____。

6. 结构的默认值是通过将所有值类型字段设置为它们的默认值，并将所有的引用类型字段设置为_____。

7. 所有的枚举默认都继承于_____。

8. 下列程序段执行后，a[4]的值为_____。

```
int []a = {1,2,3,4,5};a[4] = a[a[2]];
```

9. 要定义一个 3 行 4 列的单精度型二维数组 f，使用的定义语句为_____。

10. C# 中的字符串有两类，规则字符串和逐字字符串，定义逐字字符串时，应在其前面加上_____号。

11. 在 Array 类中，可以对一维数组中的元素进行排序的方法是_____。

12. 在 C# 中，下列代码的运行结果是_____。

```
int []price = new int[]{1,2,3,4,5,6};
foreach(int p in price)
{
    if(p%2 == 0)
        Console.Write(p);
}
```

三、阅读程序题

1.

```
using System;
namespace chp04_06_01
{
    struct MyStruct
    {
        public int x;
        public int y;
        public MyStruct(int i, int j)
        {
            x = i;
            y = j;
        }
        public void Sum()
        {
            int sum = x + y;
            Console.WriteLine("the sum is {0}", sum);
        }
    }
    class Program
    {
        static void Main(string[] args)
        {
            MyStruct s1 = new MyStruct(1, 2);
            s1.x = 2;
            s1.Sum();
        }
    }
}
```

2.

```
using System;
using System.Collections;
namespace chp04_06_02
{
    class Program
    {
        public interface Observer
        {
            void Response();         //观察者的响应,如同老鼠见到猫的反应
        }
        public interface Subject
        {
            void AimAt(Observer obs);   //针对哪些观察者,这里指猫要捕捉的对象——老鼠
        }
        public class Mouse : Observer
        {
            private string name;
```

```csharp
        public Mouse(string name, Subject subj)
        {
            this.name = name;
            subj.AimAt(this);
        }
        public void Response()
        {
            Console.WriteLine(name + " attempt to escape!");
        }
    }
    public class Master : Observer
    {
        public Master(Subject subj)
        {
            subj.AimAt(this);
        }

        public void Response()
        {
            Console.WriteLine("Host waken!");
        }
    }
    public class Cat : Subject
    {
        private ArrayList observers;
        public Cat()
        {
            this.observers = new ArrayList();
        }
        public void AimAt(Observer obs)
        {
            this.observers.Add(obs);
        }
        public void Cry()
        {
            Console.WriteLine("Cat cryed!");
            foreach (Observer obs in this.observers)
            {
                obs.Response();
            }
        }
    }
    class MainClass
    {
        static void Main(string[] args)
        {
            Cat cat = new Cat();
            Mouse mouse1 = new Mouse("mouse1", cat);
            Mouse mouse2 = new Mouse("mouse2", cat);
            Master master = new Master(cat);
            cat.Cry();
```

            }
        }
    }
}

3.

```
using System;
namespace chp04_06_03
{
    class Program
    {
        static void Main(string[] args)
        {
            MyClass m = new MyClass();
            int[] s = { 34, 23, 65, 67, 54, 98, 6, 56 };
            m.Array(s);
            for (int i = 0; i < s.Length; i++)
                Console.Write("{0}", s[i]);
        }
    }
    class MyClass
    {
        public void Array(int[] a)
        {
            for (int i = 0; i < a.Length; i++)
                a[i] = i;
        }
    }
}
```

4.

```
using System;
namespace chp04_06_04
{
    class Test
    {
        static int[] a = { 1, 2, 3, 4, 5, 6, 7, 8 };
        public static void Main()
        {
            int s0, s1, s2;
            s0 = s1 = s2 = 0;
            for (int i = 0; i < 8; i++)
            {
                switch (a[i] % 3)
                {
                    case 0: s0 += Test.a[i]; break;
                    case 1: s1 += Test.a[i]; break;
                    case 2: s2 += Test.a[i]; break;
                }
            }
```

```csharp
            Console.WriteLine(s0 + " " + s1 + " " + s2);
        }
    }
}
```

5.

```csharp
using System;
namespace chp04_06_05
{
    class Program
    {
        static void Main(string[] args)
        {
            int[] myArray = { 115, 150, 198, 30,28 };
            for (int i = 0; i != myArray.Length - 1; i++)
                AddArray(myArray[i]);
            foreach (int i in myArray)
                Console.Write("{0}   ",i);
        }
        public static void AddArray(int num)
        { num += 1; }
    }
}
```

6.

```csharp
using System;
namespace chp04_06_06
{
    class Program
    {
        private class IntIndexer
        {
            private string[] myData;
            public IntIndexer(int index)
            {
                myData = new string[index];
                for (int i = 0; i < index; i++)
                    myData[i] = "Microsoft";
            }
            public string this[int pos]
            {
                get
                {
                    return myData[pos];
                }
                set
                {
                    myData[pos] = value;
                }
            }
```

```
        }
        static void Main(string[] args)
        {
            int size = 3;
            IntIndexer myIndex = new IntIndexer(size);
            myIndex[2] = "Visual";
            myIndex[1] = "Studio";
            for (int i = 0; i < size; i++)
                Console.Write(myIndex[i] + " ");
        }
    }
}
```

四、程序填空题

1. 下面的程序是用来在数组 table 中查找 x，若数组中存在 x，程序输出数组中第一个等于 x 的数组元素的下标，否则输出 −1。

```
using System;
using System.Collections.Generic;
using System.Linq;
using System.Text;
using System.Threading.Tasks;
namespace chp04_06_07
{
    class Program
    {
        static void lookup(int[] t, ref int i, int val, int n)
        {
            int k;
            for (k = 0; k < n; k++)
                if (_____①_____) { i = k; return; }
            _____②_____ ;
            return;
        }
        static void Main(string[] args)
        {
            int[] table = { 12, 34, 54, 23, 45, 33, 78, 87, 59, 97 };
            int x, index = 0;
            x = int.Parse(Console.ReadLine());
            lookup(table, _____③_____ , x, 10);
            Console.WriteLine(" {0}", index);
        }
    }
}
```

2. 下面程序实现对数组 array 的冒泡排序。

```
using System;
namespace chp04_06_08
{
    class Program
    {
        static void Main(string[] args)
```

```
            {
                int[ ] array = { 20, 56, 38, 45 };
                int temp;
                for (int i = 0; i < 3; i++)
                    for (int j = 0; j <____①____; j++)
                        if (a[j] < a[j + 1])
                        {
                            temp = a[j];
                            array[j] = ____②____ ;
                            array[j + 1] = temp;
                        }
                for(int i = 0;i < = 3;i++)
                    Console.WriteLine(array[i]);
            }
        }
    }
```

3. 下面程序输出数组元素中的最大值和最小值。

```
using System;
namespace chp04_06_09
{
    class Program
    {
        static void Main(string[ ] args)
        {
            int m, n, i;
            int[ ] a = new int[6]{1,2,5,3,9,7};
            m = n = a[0];
            for (i = 1; i < 6; i++)
            {
                if (____①____) m = a[i];
                if (____②____) n = a[i];
            }
            Console.WriteLine("最大值 = {0},最小值 = {1}",m,n );
            Console.ReadLine();
        }
    }
}
```

4. 键盘输入 10 个数值,统计输出的平均数及大于等于平均数的数值个数。请完善程序。

```
using System;
namespace chp04_06_10
{
    class Program
    {
        static void Main(string[ ] args)
        {
```

```
            int[] a = new int[10];
            int sum = 0,    ①    ;
            double avg;
            for (int i = 0; i < 10; i++)
            {
                a[i] = int.Parse(Console.ReadLine());
                    ②    ;
            }
            avg = sum / 10.0;
            Console.WriteLine("平均值是{0}", avg);
            for (int i = 0; i < 10; i++)
                if (    ③    )
                    n++;
            Console.WriteLine("大于平均数的数值个数是{0}", n);
        }
    }
}
```

五、编写程序题

1. 编写程序，统计 4×5 二维数组中奇数的个数和偶数的个数。

2. 定义一个行数和列数相等的二维数组，并执行初始化，然后计算该数组两条对角线上的元素值之和。写一个函数在给定的字符串中查找指定的字符。若找到，返回该字符的地址；否则，返回 NULL 值。然后调用该函数输出一字符串中从指定字符开始的全部字符，如输入 abcbde 和 b，则输出 bcbde。

3. 定义一个描述 3 种颜色的枚举类型（red、blue、green），输出这 3 种颜色的全部排列结果。

4. 从键盘上输入 3 个正整数，分别表示某年某月某日，计算它们对应于该年的第多少天，并输出结果值。要求年、月、日均作为某结构体的成员。

5. 统计候选人得票的程序。设有 3 个候选人，最终只能有 1 人当选。现有 10 人参加投票，从键盘先后输入这 10 人所投候选人的名字，要求输出 3 名候选人的得票结果。

习题 7　Windows 窗体与控件

一、选择题

1. 当运行程序时，系统自动执行启动窗体的（　　）事件。
　　A. Click　　　　　B. DoubleClick　　C. Load　　　　　D. Activated
2. 若要使命令按钮不可操作，要对（　　）属性进行设置。
　　A. Visible　　　　B. Enabled　　　　C. BackColor　　　D. Text
3. 若要使 TextBox 中的文字不能被修改，应对（　　）属性进行设置。
　　A. Locked　　　　B. Visible　　　　C. Enabled　　　　D. ReadOnly
4. 以下不是构造函数特征的是（　　）。
　　A. 构造函数的函数名与类名相同　　　B. 构造函数可以重载
　　C. 构造函数可以设置默认参数　　　　D. 构造函数必须指定类型说明

5. 在设计窗口,可以通过( )属性向列表框控件(如 ListBox)的列表添加项。
   A. Items   B. Items.Count   C. Text   D. SelectedIndex

6. 引用 ListBox(列表框)最后一个数据项应使用( )语句。
   A. ListBox1.Items[ListBox1.Items.Count]
   B. ListBox1.Items[ListBox1.SelectedIndex]
   C. ListBox1.Items[ListBox1.Items.Count－1]
   D. ListBox1.Items[ListBox1.SelectedIndex－1]

7. 引用 ListBox(列表框)当前被选中的数据项应使用( )语句。
   A. ListBox1.Items[ListBox1.Items.Count]
   B. ListBox1.Items[ListBox1.SelectedIndex]
   C. ListBox1.Items[ListBox1.Items.Count－1]
   D. ListBox1.Items[ListBox1.SelectedIndex－1]

8. 窗体中有一个年龄文本框 txtAge,下面( )代码可以获得文本框中的年龄值。
   A. int age = txtAge;
   B. int age = txtAge.Text;
   C. int age = Convert.ToInt32(txtAge);
   D. int age = int.Parse(txtAge.Text);

9. 下面( )代码可以显示一个消息框。
   A. Dialog.Show();   B. MessageBox.Show();
   C. Form.Show();   D. Form.ShowDialog();

10. ( )可以构建 Windows 窗体以及其所使用空间的所有类的命名空间。
    A. System.IO   B. System.Data
    C. System.Text   D. System.Windows.Forms

11. ( )控件组合了 TextBox 控件和 ListBox 控件的功能。
    A. ComboBox   B. Label
    C. ListView   D. DomainUpDown

12. 和 C#中的所有对象一样,窗体也是对象,是( )类的实例。
    A. Label   B. Controls   C. Form   D. System

13. 在 C#程序中为显示变量 mainForm 引用的窗体对象,必须( )。
    A. 使用 mainForm.ShowDialog()方法
    B. 将 mainForm 对象的 isDialog 属性设置为 true
    C. 将 mainForm 对象的 FromBorderStyle 枚举属性设置为 FixedDialog
    D. 将变量 mainForm 改为引用 System.Windows.Dialog 类的对象

14. 在 C#程序中,文本框控件的( )属性用来设置其是否为只读的。
    A. ReadOnly   B. Locked   C. Lock   D. Style

15. 在 C#程序中,显示一个信息为"This is a test!",标题为 Hello 的消息框,正确的语句是( )。
    A. MessageBox("this is a test!","Hello");
    B. MessageBox.Show("this is a test!","Hello");

C. MessageBox("Hello","this is a test!");

D. MessageBox.Show("Hello","this is a test!");

16. 消息对话框不是放置到窗体上的,是使用 MessageBox 的( )方法显示出来的。
    A. Move   B. Show   C. Control   D. Load

17. 在 Windows 应用程序中,如果复选框控件的 Checked 属性值设置为 true,表示( )。
    A. 该复选框被选中   B. 该复选框不被选中
    C. 不显示该复选框的文本信息   D. 显示该复选框的文本信息

18. 要获取 ComboBox 控件所包含项的集合,可以通过( )属性得到。
    A. SelectedItem   B. SelectedText
    C. Items   D. Sorted

19. 启动一个定时器控件的方法是( )。
    A. Enabled   B. Interval   C. Start   D. Stop

20. 要改变 Button 控件上显示的文本内容,应修改( )属性。
    A. Name   B. Tag   C. Text   D. Font

二、填空题

1. Timer 控件的_____属性,用来指定时钟控件触发时间的时间间隔,单位 ms。

2. _____属性用于获取 ListBox1 控件中项的数目。

3. ComboBox 控件的 SelectedIndex 属性返回对应于组合框中选定项的索引整数值,其中,第 1 项为_____,未选中为_____。

4. 要使 Lable 控件显示给定的文字"您好",应设置它的_____属性值。

5. 在 Visual Studio 中双击窗体中的某个按钮,则会自动添加该按钮的_____事件。

6. 实现密码框功能的方法是将 TextBox 控件的_____属性赋予屏蔽字符。

7. 当进入 Visual Studio 集成环境,如果没有显示"工具箱"窗口,应选择"_____"菜单项的"工具箱"选项,以显示"工具箱"窗口。

8. 要让某控件不可用,应该设置该控件的_____属性,使其值为_____。

9. 要让某控件不可见,应该设置该控件的_____属性,使其值为_____。

10. PictureBox 控件可以显示不同格式的图像文件,通过_____属性将图像文件装载显示在控件框内,还可以通过设置_____属性设置处理图像大小和控件的关系。如要将图像的大小自动缩放和 PixtureBox 控件相适应,则应将该属性设置为_____模式。

11. 要取出 ProgressBar 控件的当前值,可以通过读取_____属性值得到。

12. 编辑带格式的文本,可以使用_____控件。

13. 往 ComboBox1 控件的选项中添加一个项目"长沙"的语句是_____。

14. ListBox 控件的_____属性可以设置列表框控件的列表选择模式。

15. 如果需要在一个日历表中选中某一天,则显示该天属于哪一个星座,应该在 MonthCalendar 控件的_____事件中编写代码。

三、阅读程序题

1. 程序运行界面如图 4-2 所示，请写出此时单击 button1 按钮后的程序运行结果。

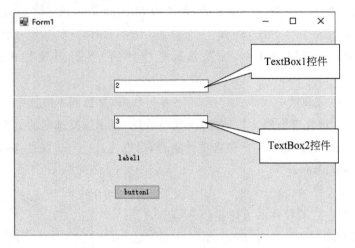

图 4-2　文本框的应用程序运行界面

程序代码如下：

```csharp
using System;
using System.Windows.Forms;
namespace chp04_07_01
{
    public partial class Form1 : Form
    {
        public Form1()
        {
            InitializeComponent();
        }
        private long fact(int k)
        {
            long f = 1;
            for (int i = 1; i <= k; i++)
                f = f * i;
            return f;
        }
        private void button1_Click_1(object sender, EventArgs e)
        {
            int a, b;
            long sum;
            a = int.Parse(textBox1.Text);
            b = int.Parse(textBox2.Text);
            sum = fact(a) + fact(b);
            label1.Text = a.ToString() + "! +" + b.ToString() +
                          "!= " + sum.ToString();
        }
    }
}
```

2. 程序运行界面如图 4-3 所示,请写出此时单击 button1 按钮后的程序运行结果。

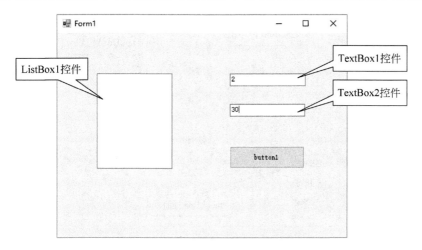

图 4-3 列表框的应用程序运行界面

程序代码如下:

```
using System;
using System.Windows.Forms;
namespace chp04_07_02
{
    public partial class Form1 : Form
    {
        public Form1()
        {
            InitializeComponent();
        }
        private bool prime(int a)
        {
            for (int i = 2; i <= Math.Sqrt(a); i++)
                if (a % i == 0)
                    return false;
            return true;
        }
        private void button1_Click(object sender, EventArgs e)
        {
            int a, b;
            a = int.Parse(textBox1.Text);
            b = int.Parse(textBox2.Text);
            for (int i = a; i <= b; i++)
                if (prime(i) == true)
                    listBox1.Items.Add(i.ToString());
        }
    }
}
```

3. 程序运行界面如图 4-4 所示,请写出此时单击 button1 按钮后的程序运行结果。

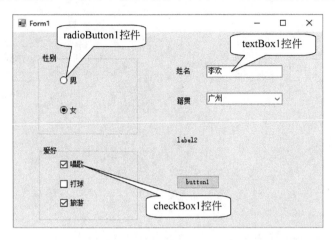

图 4-4　单选按钮和复选框应用程序运行界面

程序代码如下：

```csharp
using System;
using System.Windows.Forms;
namespace chp04_07_03
{
    public partial class Form1 : Form
    {
        public Form1()
        {
            InitializeComponent();
        }
        private void button1_Click(object sender, EventArgs e)
        {
            label2.Text = textBox1.Text;
            if (radioButton1.Checked == true)
                label2.Text = label2.Text + "性别：" + radioButton1.Text;
            else
                label2.Text = label2.Text + "性别：" + radioButton2.Text;
                label2.Text = label2.Text + "籍贯：" + comboBox1.SelectedItem + " 喜欢";
            if (checkBox1.Checked == true)
                label2.Text = label2.Text + checkBox1.Text + " ";
            if (checkBox2.Checked == true)
                label2.Text = label2.Text + checkBox2.Text + " ";
            if (checkBox3.Checked == true)
                label2.Text = label2.Text + checkBox3.Text;
        }
    }
}
```

4. 程序运行界面如图 4-5 所示，请写出此时单击 button1 按钮后的程序运行结果。

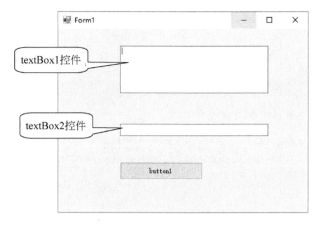

图 4-5　文本框的应用程序运行界面

程序代码如下：

```
using System;
using System.Windows.Forms;
namespace chp04_07_04
{
    public partial class Form1 : Form
    {
        public Form1()
        {
            InitializeComponent();
        }
        private void button1_Click(object sender, EventArgs e)
        {
            textBox1.Multiline = true;
            textBox1.Text = "寒雨连江夜入吴,平明送客楚山孤.\r\n洛阳亲友如相问,一片冰心在玉壶.";
            textBox1.Height = 100;
            textBox1.SelectionStart = 0;
            textBox1.SelectionLength = 7;
            textBox2.Text = textBox1.SelectedText;
        }
    }
}
```

四、程序填空题

1. Windows 应用程序实现：任意输入两个整数，当输入的字符不是数字时给出提示信息。单击按钮后将两数之间的回文数（如 121）输出到列表框里。程序界面如图 4-6 所示。请完善程序。

程序代码如下：

```
using System;
using System.Windows.Forms;
namespace chp04_07_05
{
```

```csharp
public partial class Form1 : Form
{
    public Form1()
    {
        InitializeComponent();
    }
    private void textBox1_KeyPress(object sender, KeyPressEventArgs e)
    {
        if(!char.IsDigit(e.KeyChar))
        {
            MessageBox.Show("请输入数字!", "提示",
                MessageBoxButtons.OK, MessageBoxIcon.Information);
            e.Handled = true;
        }
    }
    private void textBox2_KeyPress(object sender, KeyPressEventArgs e)
    {
        if (!char.IsDigit(e.KeyChar))
        {
            MessageBox.Show("请输入数字!", "提示",
                MessageBoxButtons.OK, MessageBoxIcon.Information);
            e.Handled = true;
        }
    }
    bool huiwen(int k)
    { int s = 0;
        for (int j = k; j != 0; j = j / 10)
            ____①____ ;
        return s == k;
    }
    private void button1_Click(object sender, EventArgs e)
    {
        for (int i = int.Parse(textBox1.Text); i <= int.Parse(textBox2.Text); i++)
            if (____②____)
                listBox1.Items.Add(____③____);
    }
}
```

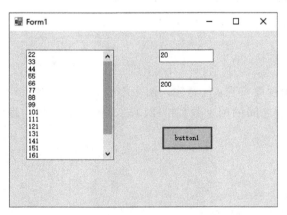

图 4-6 列表框的应用程序运行界面

2. Windows 应用程序界面如图 4-7 所示,单击 button1 按钮后将 listBox1 的数取出来送到数组 a 中,再将 a 数组按从大到小排序,并将结果送入 listBox2。请完善程序。

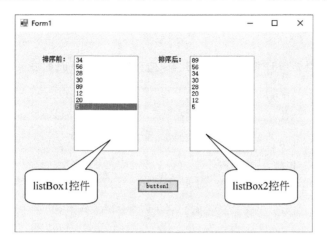

图 4-7 列表框的应用程序运行界面

程序代码如下:

```
using System;
using System.Windows.Forms;
namespace chp04_07_06
{
    public partial class Form1 : Form
    {
        public Form1()
        {
            InitializeComponent();
        }
        private void button1_Click(object sender, EventArgs e)
        {
            int[] a = new int[listBox1.Items.Count];
            for(int i = 0;     ①    ;i++)
            {
                this.listBox1.SetSelected(i, true);
                a[i] = int.Parse(this.listBox1.SelectedItem.ToString());
            }
            for(int i = 0;i <= a.Length - 2;i++)
            {
                for(int j = i + 1;    ②    ;j++)
                    if(a[i]< a[j])
                    {
                        int t;
                        t = a[i];
                        a[i] = a[j];
                        a[j] = t;
                    }
                    ③     ;
            }
```

```
            listBox2.Items.Add(a[a.Length - 1].ToString());
        }
    }
}
```

五、编写程序题

1. 设计一个模拟简单计算器的 Windows 应用程序，其界面如图 4-8 所示。

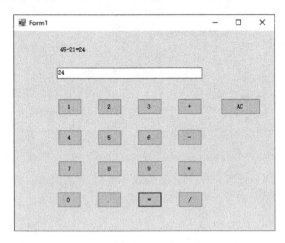

图 4-8　简单计算器程序运行界面

2. 设计一个 Windows 应用程序，利用 TextBox、ComboBox、GroupBox、RadioButton、CheckBox、RichTextBox、ListBox、DateTimePicker 等控件完成个人信息的输入，并将输入的信息汇总输出到"简介"文本框内，其界面如图 4-9 所示。

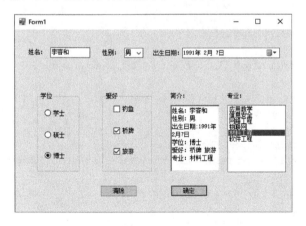

图 4-9　多控件应用程序运行界面

习题 8　用户界面设计

一、选择题

1. 创建菜单后，为了实现菜单项的命令功能，应为菜单项添加（　　）事件处理方法。
　　A. DrawItem　　　　B. Popup　　　　C. Click　　　　D. Select

2. 设置 openfiledialog1 引用一个 openfiledialog 对象,则打开该对话框的正确代码是(　　)。
　　A. openfiledialog1.show();　　　　　　B. openfiledialog1.showdialog();
　　C. openfiledialog1.open();　　　　　　D. openfiledialog1.openandshow();
3. 右击一个控件时出现的菜单一般称为(　　)。
　　A. 主菜单　　　　B. 菜单项　　　　C. 快捷菜单　　　　D. 子菜单
4. 变量 openFile1 引用一个 OpenFileDialog 对象,为检查用户在退出对话框时是否单击了"确认"按钮,应检查 openFile1.ShowDialog()的返回值是否等于(　　)。
　　A. DiaglogResult.OK　　　　　　　　B. DialogResult.Yes
　　B. DialogReslut.Cancel　　　　　　　D. DialogResult.No
5. 下面关于菜单的描述中,错误的是(　　)。
　　A. 菜单是用户界面的重要组成部分
　　B. C#的菜单按使用形式可以分为下拉式菜单和快捷菜单两种
　　C. 在设计菜单界面后,应该为菜单项编写事件过程
　　D. 菜单项唯一响应的事件就是 Click 事件
6. C#通过工具栏控件为用户提供了直观、快捷的操作方式,下列(　　)不能出现在工具栏里。
　　A. Button 控件　　　　　　　　　　　B. TreeView 控件
　　C. Label 控件　　　　　　　　　　　 D. ComboBox 控件
7. C#的状态栏控件为应用程序提供了显示当前状态的区域,状态栏可以放置的控件有(　　)。
　　A. Button 控件、Label 控件　　　　　　B. TextBox 控件、ProgressBar 控件
　　C. StatusLabel 控件、ProgressBar 控件　 D. CheckBox 控件、RadioButton 控件
8. 在打开文件对话框中,如果允许对话框选择多个文件,则应先设置(　　)属性为 true。
　　A. CheckFileExits　　　　　　　　　　B. MutiSelect
　　C. ReadOnlyChecked　　　　　　　　 D. ValiDateNames
9. C#中的颜色对话框用来在调色板中选择颜色或者创建自定义颜色,如果在打开该对话框后,限制用户只选择纯色,则需要先设置(　　)属性为 true。
　　A. AllowFullOpen　　　　　　　　　　B. AnyColor
　　C. FullOpen　　　　　　　　　　　　 D. SolidColorOnly
10. 如果要隐藏并禁用菜单项,需要设置(　　)两个属性。
　　A. Visible 和 Enable　　　　　　　　　B. Visible 和 Enabled
　　C. Visual 和 Enable　　　　　　　　　C. Visual 和 Enabled

二、填空题
1. 设计快捷菜单可以利用 C#提供的_____控件,设计好快捷菜单项后,假设该快捷菜单名为 contextMenu1,还要将它与窗口控件相关联,其方法是将该窗口控件的_____属性设置为建立好的快捷菜单_____。
2. 如果要在窗体上创建一个标准菜单,可以直接向窗体添加一个_____控件。用户

可以将_____控件理解为一个容器,它包含了所有菜单项的对象集,每个菜单项是一个_____。

3. 通过设置菜单项的_____属性,可以指示选中标记是否出现在菜单项文本的旁边。如果要放置选中标记在菜单项旁边,则该值应设为_____,否则为_____。

4. 如果需要打开文件,可以使用通用对话框的_____控件实现,假设要打开的文件类型为 rtf 文件或 txt 文件,则可设置其 Filter 属性为_____。

5. 在 C#中,可以使用_____对话框设置并返回所用字体的名称、样式、大小等属性,使用_____方法启动其对话框。

6. 如果需要指定保存文件的路径和文件名,可以使用_____通用对话框,其中_____属性用来返回指定或者选中的文件名的字符串。

三、阅读程序题

1. 阅读下面的程序代码,程序界面如图 4-10 所示,理解程序实现的功能。

图 4-10 菜单操作程序运行界面

```
using System;
using System.Windows.Forms;
namespace chp04_08_01
{
    public partial class Form1 : Form
    {
        bool expand = true;
        public Form1()
        {
            InitializeComponent();
        }
        private void 展开关闭其他项ToolStripMenuItem_Click(object sender, EventArgs e)
        {
            switch (expand)
            {
                case true:
                    this.员工录入ToolStripMenuItem.Visible = true;
                    this.修改密码ToolStripMenuItem.Visible = true;
                    this.设置密码ToolStripMenuItem.Visible = true;
                    expand = false;
                    this.操作ToolStripMenuItem.ShowDropDown();
```

```
                    break;
                case false:
                    this.员工录入ToolStripMenuItem.Visible = false;
                    this.修改密码ToolStripMenuItem.Visible = false;
                    this.设置密码ToolStripMenuItem.Visible = false;
                    expand = true;
                    this.操作ToolStripMenuItem.ShowDropDown();
                    break;
            }
        }
    }
}
```

2. 阅读下面的程序代码,程序界面如图 4-11 所示,理解程序实现的功能。

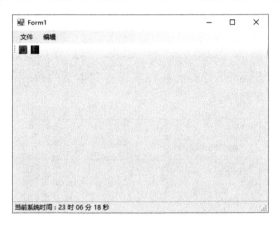

图 4-11 工具栏应用程序运行界面

向窗体中添加一个 MenuStrip 控件,用于显示菜单信息;添加一个 ToolStrip 控件,用于显示工具栏信息;添加一个 StatusStrip 控件,用于动态显示系统时间。

```
using System;
using System.Threading;
using System.Windows.Forms;
namespace chp04_08_02
{
    public partial class Form1 : Form
    {
        public Form1()
        {
            InitializeComponent();
        }
        private void Form1_Load(object sender, EventArgs e)
        {
            Thread P_th = new Thread(            //创建线程对象
            () =>                                //使用 Lambda 表达式
            {
                while (true)                     //无限循环
                {
```

```
                    Invoke(
                    (MethodInvoker)(            //通过委托调用窗体线程
                    () =>
                    {
                        toolStripStatusLabel1.Text = "当前系统时间：" +
                            DateTime.Now.ToString("HH时 mm 分 s 秒");
                    }));
                    Thread.Sleep(1000);          //线程挂起 1s
                }
            }
            );
            P_th.IsBackground = true;            //设置线程为后台线程
            P_th.Start();                        //线程开始
        }
    }
}
```

四、程序填空题

1．下面的程序利用 SaveFileDialog 控件实现将文本框的内容保存到文本文件中去，界面如图 4-12 所示。请完善程序。

图 4-12　文件保存对话框应用程序运行界面

```
using System;
using System.Windows.Forms;
namespace chp04_08_03
{
    public partial class Form1 : Form
    {
        public Form1()
        {
            InitializeComponent();
        }
        private void button1_Click(object sender, EventArgs e)
        {
            saveFileDialog1.Filter = ____①____ ;
            if(saveFileDialog1.ShowDialog() == ____②____ )
            {
                System.IO.StreamWriter myfile = new
```

```
                    System.IO.StreamWriter(    ③    );
            myfile.Write(textBox1.Text);
                ④        ;
        }
    }
  }
}
```

2. 下面的程序利用 FontDialog 和 ColorDialog 设置文本框的字体和颜色,界面如图 4-13 所示。请完善程序。

图 4-13 字体和颜色设置对话框应用程序运行界面

```
using System;
using System.Windows.Forms;
namespace chp04_08_04
{
    public partial class Form1 : Form
    {
        public Form1()
        {
            InitializeComponent();
        }
        private void button1_Click(object sender, EventArgs e)
        {
            if(this.fontDialog1.ShowDialog() ==    ①    )
                this.textBox1.Font =    ②    ;
        }
        private void button2_Click(object sender, EventArgs e)
        {
            if(this.colorDialog1.ShowDialog() ==    ③    )
                textBox1.ForeColor =    ④    ;
        }
    }
}
```

五、编写程序题

1. 定义一个结构数组用来存放 8 个学生的记录,设计一个带工具栏的 Windows 窗体应用程序,可以通过工具栏的按钮浏览结构数组中的记录。窗口界面如图 4-14 所示,设计程序完成其功能。

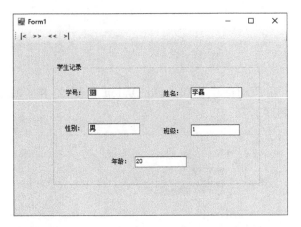

图 4-14　工具栏应用程序运行界面

2. 编写一个简易的文本编辑器,具备打开、保存文本文件,对所选文本设置字体、字号、字型及颜色等功能,分别用菜单和工具栏方式提供其功能列表。窗口界面如图 4-15 所示,设计程序完成其功能。

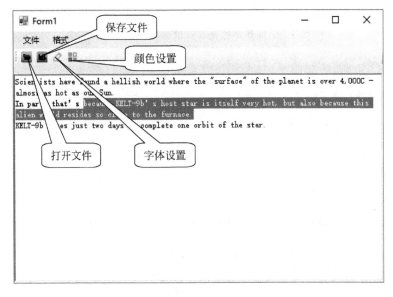

图 4-15　基本对话框应用程序运行界面

习题 9　文 件 操 作

一、选择题

1. 在使用 FileStream 打开一个文件时,通过使用 FileMode 枚举类型的(　　)成员,来指定操作系统打开一个现有文件并把文件读写指针定位在文件尾部。

A. Append   B. Create   C. CreateNew   D. Truncate

2. 指定操作系统读取文件方式中 FileMode.Create 的含义是(　　)。

　A. 打开现有文件

　B. 指定操作系统应创建文件,如果文件存在,将出现异常

　C. 打开现有文件,若文件不存在,出现异常

　D. 指定操作系统应创建文件,如果文件存在,将被改写

3. 在一个 C# 的 Console 应用程序中,Main 函数的执行代码如下:

```
static void Main(string [] args)
{
    Console.WriteLine("请为文件输入一个名称: ");
    string Filename = Console.ReadLine();
    FileStream filestr = new FileStream(Filename, FileMode.OpenOrCreate);
    BinaryWriter binwrt = new BinaryWriter(filestr);
    for (int i = 0; i < 20; i++)
    {
        binwrt.Write( (int) i);
    }
    filestr.Close();
}
```

以上代码行存在问题的是(　　)。

　A. 没有指定文件所在的目录,所以不能创建文件

　B. binwrt 对象不能写入整数数据,只能写入字符数据,因此将提示错误信息

　C. binwrt 对象在执行完毕后应调用 Close 方法关闭

　D. 上述代码没有问题,将正常执行

4. `FileStream fs = new Filestream`
　　`("c:\\test.txt",FileMode.Create,FileAccess.ReadWrite,Fileshare.None);`

针对如上 C# 代码段,以下说法正确的是(　　)。

Ⅰ. 如果 c 盘根目录下已经存在文件 test.txt,则编译报错

Ⅱ. 如果 c 盘根目录下已经存在文件 test.txt,则改写 test.txt 文件,将其内容清空

Ⅲ. 如果 c 盘根目录下已经存在文件 test.txt,则不做任何操作,但对该文件持有读写权

Ⅳ. 如果 c 盘根目录下不存在文件 test.txt,则建立一个内容为空的 test.txt 文件

　A. Ⅰ和Ⅱ   B. Ⅱ和Ⅲ   C. Ⅲ和Ⅳ   D. Ⅱ和Ⅳ

5. 以下的 C# 代码用来读写特定的文件:

```
using System;
using System.IO;
public class FileReader
{
    public static void Main(string[] args)
    {
        string Filename = "ACCP.TXT";
        if(!File.Exists(Filename))
        {
```

```
            Console.WriteLine("{0} 不存在!", Filename);
            return;
        }
        StreamReader sr = File.OpenText(Filename);
        string input;
        while((input = sr.ReadLine()) != null)
            Console.WriteLine(input);
        Console.WriteLine();
        sr.Close();
    }
}
```

假设当前文件下的文件 ACCP.TXT 的内容有两行并且内容为：

AAAAA
BBBBB

则以下说法正确的是（　　）。

  A. 程序中存在错误，因为代码行 StreamReader sr = File.OpenText(Filename); 必须指定文件打开的模式

  B. 程序将打印输出"ACCP.TXT 不存在!"并退出

  C. 程序无错误，输出的数据行为一行

  D. 程序无错误，并且最后输出两行数据

6. 在.NET 中，以下都是 System.IO 中定义的类，除了（　　）。

  A. TextWriter   B. Stream

  C. FileReader   D. FileSystemInfo

7. 以下不属于文件访问方式的是（　　）。

  A. 读/写   B. 只读   C. 只写   D. 不读不写

8. 以下（　　）类提供了文件夹的操作功能。

  A. File   B. Directory   C. FileStream   D. BinaryWriter

9. 用 FileStream 对象打开一个文件时，可以用 FileMode 参数控制（　　）。

  A. 对文件进行只读、只写还是读/写操作

  B. 对文件覆盖、创建、打开等选项中的哪些操作

  C. 对文件进行随机访问时的偏移位置

  D. 其他 FileStream 对象对同一个文件所具有的访问类型

10. FileStream 类的（　　）方法用来从流中读取字节块，并将该数据写入到基础设备中。

  A. EndRead   B. ReadByte   C. Read   D. BeginRead

11. （　　）是字节序列的抽象概念，是一种向存储设备写入字节和从存储设备读取字节的方式。

  A. File   B. Stream   C. Buffer   D. ArrayList

12. Directory 类的（　　）方法用于创建指定路径中包含的所有目录和子目录并返回一个 DirectoryInfo 对象，通过该对象操作相应的目录。

  A. CreateDirectory( )   B. Path( )

  C. Create( )   D. Directory( )

二、填空题

1. 文件是存储在磁盘等介质上的一组数据，按文件的存取方式及结构，文件可以分为顺序文件和_____。按文件数据的组织形式，文件可分为 ASCII 文件和_____。

2. System.IO 模型是一个文件操作类库，其中_____类提供了可用于创建文件_____、复制_____、删除_____和打开文件的静态方法。

3. _____类提供了文件夹操作的方法，类内的方法是静态的，无须创建对象即可使用，其中_____方法可以创建文件夹，_____方法可以用来删除文件夹。

4. 在 System.IO 模型中，文件操作的基本方式是用 File 类打开操作系统文件，建立对应的文件流，即_____对象，用_____类提供的方法对该文件流（文本文件）进行读操作，用_____类提供的方法对该文件流（文本文件）进行写操作。

5. 在文件流对象中，对二进制文件进行读操作的方法是_____，对二进制文件进行写操作的方法是_____。

6. 在 FileStream 类的使用过程中，可以使用 FileStream 类的_____属性来获取流的长度（以字节为单位），使用_____属性来获取当前读写指针的位置。

7. 在.NET 框架中，与基本输入/输出操作相关的类都位于_____，命名空间中，所以用户要在代码中使用_____语句来导入这个命名空间。

8. File 类中，_____方法以只读的方式打开指定的文件并返回 FileStream 对象。

9. 在读取数据之前，可以用 StreamReader 类的_____方法来检测是否到达了流的末尾。该函数返回流的当前位置上的字符，但不移动指针，如果到达末尾，则返回_____。

10. StreamWriter 类除了 Write 方法以外，还提供了_____方法，其用法与 Write 方法类似，只是它在写入指定数据后自动添加一个换行符。

三、阅读程序题

1. 分析以下程序，写出运行结果（假设在 E 盘已经建立了一个 data 空文件夹）。

```
using System;
using System.IO;
namespace chp04_09_01
{
    class Program
    {
        static void Main(string[] args)
        {
            string mypath = @"e:\data\test.txt";
            if (!File.Exists(mypath))
            {
                FileStream myfile = File.Create(mypath);
                if (File.Exists(mypath))
                {
                    Console.WriteLine("File '{0}' has created successful!", mypath);
                    Console.WriteLine("Folder name:" + myfile.Name);
                    Console.WriteLine("size in bytes {0}", myfile.Length);
                    Console.WriteLine("Lets use this file now.");
                }
                else
```

```csharp
            Console.WriteLine("File {0} creat failed", mypath);
                myfile.Close();
            }
            File.Delete(mypath);
            Console.WriteLine("{0} was successfully delelted.", mypath);
        }
    }
}
```

2. 分析以下程序，写出运行结果（分别讨论当 E 盘中存在 data 文件夹和不存在 data 文件夹时的输出结果）。

```csharp
using System;
using System.IO;
namespace chp04_09_02
{
    class mystream
    {
        public static void read(string myfile)
        {
            try
            {
                StreamReader fr = new StreamReader(myfile);
                string text;
                while ((text = fr.ReadLine()) != null)
                    Console.WriteLine(text);
                fr.Close();
            }
            catch (Exception e)
            {
                Console.WriteLine("The file can't be read");
                Console.WriteLine(e.Message);
            }
        }
        public static void write(string myfile)
        {
            try
            {
                StreamWriter fw = new StreamWriter(myfile);
                fw.WriteLine("Hello,World!");
                fw.WriteLine("Hello,C#");
                fw.Close();
            }
            catch (Exception e)
            {
                Console.WriteLine("The file can't be written");
                Console.WriteLine(e.Message);
            }
        }
    }
    class Program
```

```
        {
            static void Main(string[] args)
            {
                string myfile = @"e:\data\text1.txt";
                mystream.write(myfile);
                mystream.read(myfile);
            }
        }
    }
```

3. 分析以下程序,写出运行结果。

```
using System;
using System.IO;
namespace chp04_09_03
{
    class Program
    {
        static void Main(string[] args)
        {
            try
            {
                FileStream myfile = new FileStream(@"e:\data\mytext.txt", FileMode.
                                    OpenOrCreate, FileAccess.ReadWrite);
                byte[] bdata = new byte[26];
                for (int i = 0; i < 26; i++)
                    bdata[i] = (byte)('a' + i);
                myfile.Write(bdata, 0, bdata.Length);
                myfile.Position = 0;
                for (int i = 0; i < myfile.Length; i++)
                    Console.Write((char)myfile.ReadByte());
                myfile.Close();
            }
            catch(Exception ex)
            {
                Console.WriteLine(ex.Message);
            }
        }
    }
}
```

四、程序填空题

1. 下面的程序使用 FileInfo 类实现了对文件的读、写、复制和删除的操作。请完善程序。

```
using System;
using System.IO;
namespace chp04_09_04
{
    class myfileinfo
    {
```

```csharp
        public string mypath1 = @"E:\data\text001.txt";
        public string mypath2 = @"E:\data\text002.txt";
        public void writefile(string str)
        {
            FileInfo file1 = _____①_____;
            if (!file1.Exists)
            {
                StreamWriter wstream1 = _____②_____;
                wstream1.WriteLine(str);
                wstream1.Close();
            }
        }
        public void readfile(string strfile, out string str)
        {
            FileInfo file2 = new FileInfo(strfile);
            str = "";
            if (_____③_____)
            {
                StreamReader rstream1 = file2.OpenText();
                string s = "";
                while ((s = rstream1.ReadLine()) != null)
                    str = str + s + "\r\n";
                rstream1.Close();
            }
            else
                Console.WriteLine("source file not exists");
        }
        public void copyfile(string path1)
        {
            FileInfo file1 = new FileInfo(mypath1);
            if (file1.Exists)
                file1.CopyTo(_____④_____);
            else
                Console.WriteLine("source file not exists!");
        }
        public void deltefile(string path1)
        {
            FileInfo file1 = new FileInfo(path1);
            if (file1.Exists)
                file1.Delete();
            else
                Console.WriteLine("file does not exists");
        }
    }
    class Program
    {
        static void Main(string[] args)
        {
            myfileinfo myfile = new myfileinfo();
            string str = "", str1, outstr = "";
            for (int i = 0; i <= 2; i++)
```

```
            {
                str1 = Console.ReadLine();
                str = str + str1 + "\r\n";
            }
            myfile.writefile(str);
            myfile.readfile(out outstr);
            Console.WriteLine(outstr);
            myfile.copyfile(myfile.mypath2);
            myfile.deltefile(myfile.mypath1);
        }
    }
}
```

2. 下面的程序实现将 e:\data 文件夹下的 t0.txt 和 t1.txt 合并后写入到 t2.txt。请完善程序。

```
using System.IO;
namespace chp04_09_05
{
    class Program
    {
        static void Main(string[] args)
        {
            string strline;
            StreamWriter wstream = new      ①      ;
            StreamReader rstream0 = new StreamReader(@"e:\data\t0.txt");
            strline = rstream0.ReadLine();
            while(strline!= null)
            {
                wstream.WriteLine(strline);
                    ②    ;
            }
            rstream0.Close();
            StreamReader rstream1 = new StreamReader(@"e:\data\t1.txt");
            strline = rstream1.ReadLine();
            while(    ③    )
            {
                wstream.WriteLine(strline);
                strline = rstream1.ReadLine();
            }
            rstream1.Close();
                ④    ;
        }
    }
}
```

五、编写程序题

1. 使用 File 类实现对 E:\data\text01.txt 文件的写入、读出、复制(到 E:\data\text02.txt)、删除(E:\data\text01.txt)的操作。

2. 编程实现资源管理器中显示磁盘目录结构及文件列表的功能。

习题10 图形与图像处理

一、选择题

1. GDI+是微软公司在 Windows 2000 以后操作系统中提供的新的图形设备接口,其通过一套部署为托管代码的类来展现,这套类被称为 GDI+的"托管类接口",程序员通过它可以创建(   )。

   Ⅰ. 图形对象   Ⅱ. 文本对象   Ⅲ. 图像对象

   A. Ⅰ和Ⅱ   B. Ⅱ和Ⅲ   C. Ⅰ和Ⅲ   D. Ⅰ、Ⅱ和Ⅲ

2. 在 GDI+的所有类中,(   )类是核心,它封装了 GDI+绘图面,提供将对象绘制到显示设备的方法,在绘制任何图形之前必须先用它创建一个对象。

   A. Pen   B. Font   C. Graphics   D. Brush

3. 下面创建 Graphics 对象的方法中,(   )是错误的。

   A. private void form1_Paint(object sender, PaintEventArgs e)
        { Graphics g = e.Graphics; }
   B. Graphics g = new Graphics();
   C. Grahpics g = this.CreateGraphics(); (this 为控件或窗体)
   D. Image img = Image.FromFile("p1.jpg");
        Graphics g = Graphics.FromImage(img);

4. Graphics 类包含在(   )命名空间中。

   A. System.Drawing   B. System
   C. System.Winforms   D. System.Drawing.Text

5. 下面(   )类可以结合 Graphics 对象绘制线条、勾勒形状轮廓或呈现其他几何形状。

   A. Brush   B. Font   C. Color   D. Pen

6. Graphics 类提供的绘制矩形的方法是(   )。

   A. DrawLine   B. DrawRectangle
   C. DrawArc   D. DrawPolygon

7. Graphics 类提供的绘制填充椭圆的方法是(   )。

   A. FillClosedCurve   B. FillPolygon
   C. DrawEllipse   D. FillEllipse

8. 如果要设置 Pen 对象绘制线条的宽度,应使用它的(   )属性。

   A. Color   B. Width   C. PenType   D. DashStyle

9. 在图形画布上绘制文本使用 Graphics 对象的(   )方法。

   A. DrawPie   B. FillPie   C. DrawString   D. FillPath

10. 当在 Graphics 对象上绘图完成后,需要重新绘制新的图形,这时需要清理画布对象,可以调用该画布的(　　　)法完成。

A. Clear　　　　　B. Dispose　　　　C. Restore　　　　D. SetClip

二、填空题

1. 可以通过调用当前窗体的_____方法获取对 Graphics 对象的引用,该方法把当前窗体的画笔、字体和颜色作为默认值。

2. Pen 类用于绘制线条或呈现其他几何形状,根据需要可以对 Pen 的属性进行设置,要改变画笔的颜色,可以设置 Pen 的_____属性,而_____属性可以设置 Pen 的线条样式。

3. 已经定义好 Graphics 对象 g1,在起始坐标(10,20),宽度为 100,高度为 50 的矩形框内绘制一蓝色椭圆,其语句为_____。

4. 要创建一个颜色为红色,宽度为 5 的画笔 mypen,其语句为_____。

5. 画笔用来绘制图形的边框和轮廓,若要填充图形的内部则必须使用_____对象。

6. 实心画刷指定了填充区域的颜色,定义一个颜色为绿色的实心画刷,可以用语句_____实现。

7. _____类可以用来定义文字的字体、大小和样式,创建一个字体为"宋体"、大小为 24、样式为斜体的对象,其语句是_____。

8. HatchBrush 画刷可绘制阴影效果,可以通过设置该对象的_____属性来得到不同的阴影效果。

9. 用 GDI+ 显示图像的方法是:创建 Image 类的派生类(如 Bitmap)的一个对象,再创建一个 Graphics 对象(表示要使用的绘图画布),然后调用 Graphics 对象的_____方法,将在图形类所表示的画布上绘制图像。

三、阅读程序题

1. 阅读下面的程序,描述其功能。

```
using System;
using System.Drawing;
using System.Windows.Forms;
namespace chp04_10_01
{
    public partial class Form1 : Form
    {
        public Form1()
        {
            InitializeComponent();
        }
        private void button1_Click(object sender, EventArgs e)
        {
            Graphics mydraw = this.CreateGraphics();
            Pen mypen = new Pen(Color.Red, 5);
            float startx = 400.0F;
```

```
                float starty = 200.0F;
                float controlX1 = 200.0F;
                float controlY1 = 21.0F;
                float controlX2 = 100.0F;
                float controlY2 = 10.0F;
                float endX = 100.0F;
                float endY = 50.0F;
                mydraw.DrawBezier(mypen, startx, starty, controlX1, controlY1, controlX2, controlY2,
                endX, endY);
            }
        }
    }
```

2. 阅读下面的程序,描述每个按钮实现的功能,程序运行界面如图 4-16 所示。

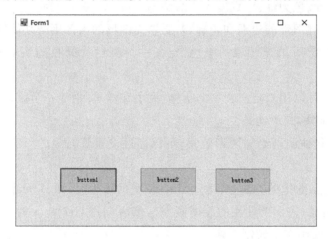

图 4-16 基本图形绘制程序运行界面

```
using System;
using System.Drawing;
using System.Drawing.Drawing2D;
using System.Windows.Forms;
namespace chp04_10_02
{
    public partial class Form1 : Form
    {
        public Form1()
        {
            InitializeComponent();
        }
        private void button1_Click(object sender, EventArgs e)
        {
            Graphics mygraphics = panel1.CreateGraphics();
            mygraphics.Clear(Color.White);
            Brush brush_back = Brushes.Red;
```

```csharp
        Brush brush_fore = Brushes.Pink;
        Font myfont = new Font("黑体", 40, FontStyle.Bold);
        string var_str = "C#程序设计基础";
        SizeF var_size = mygraphics.MeasureString(var_str, myfont);
        int var_X = (panel1.Width - Convert.ToInt32(var_size.Width)) / 2;
        int var_Y = (panel1.Height - Convert.ToInt32(var_size.Height)) / 2;
        mygraphics.DrawString(var_str, myfont, brush_back, var_X + 3, var_Y + 2);
        mygraphics.DrawString(var_str, myfont, brush_fore, var_X , var_Y );
    }
    private void button2_Click(object sender, EventArgs e)
    {
        Graphics mygraphics = panel1.CreateGraphics();
        mygraphics.Clear(Color.White);
        Brush brush_back = Brushes.Gray;
        Brush brush_fore = Brushes.Aquamarine;
        Font myfont = new Font("宋体", 40);
        string var_str = "C#程序设计基础";
        SizeF var_size = mygraphics.MeasureString(var_str, myfont);
        int var_X = (panel1.Width - Convert.ToInt32(var_size.Width)) / 2;
        int var_Y = (panel1.Height - Convert.ToInt32(var_size.Height)) / 2;
        mygraphics.TranslateTransform(var_X, var_Y);
        Matrix Var_trans = mygraphics.Transform;
        Var_trans.Shear(-0.8F, 0.00F);
        mygraphics.Transform = Var_trans;
        mygraphics.DrawString(var_str, myfont, brush_back, 25,5);
    }
    private void button3_Click(object sender, EventArgs e)
    {
        Graphics mygraphics = panel1.CreateGraphics();
        mygraphics.Clear(Color.White);
        Color color_up = Color.Red;
        Color color_down = Color.Yellow;
        Font myfont = new Font("宋体", 40);
        string var_str = "C#程序设计基础";
        SizeF var_size = mygraphics.MeasureString(var_str, myfont);
        PointF var_point = new PointF(15, 25);
        RectangleF var_rect = new RectangleF(var_point, var_size);
        LinearGradientBrush var_lbrush = new LinearGradientBrush(var_rect,
            color_up,color_down, LinearGradientMode.BackwardDiagonal);
        mygraphics.DrawString(var_str, myfont, var_lbrush, var_point);
    }
}
```

四、程序填空题

1. 下面的程序运行时,单击"绘图"按钮后的界面如图 4-17 所示。请完善程序。

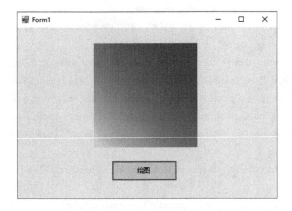

图 4-17　颜色填充应用程序运行界面

```
using System;
using System.Drawing;
using System.Drawing.Drawing2D;
using System.Windows.Forms;
namespace chp04_10_03
{
    public partial class Form1 : Form
    {
        public Form1()
        {
            InitializeComponent();
        }

        private void button1_Click(object sender, EventArgs e)
        {
            Graphics g =　    ①    ;
            Rectangle rect = new Rectangle(150,30, 200, 200);
            LinearGradientBrush lBrush = new LinearGradientBrush(rect,
                Color.Red, Color.Yellow, LinearGradientMode.BackwardDiagonal);
                　②　;
        }
    }
}
```

2. 下面的程序运行时，单击"绘图"按钮后的界面如图 4-18 所示。请完善程序。

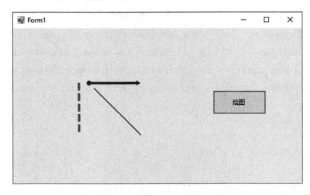

图 4-18　基本图形绘制程序运行界面

```
using System;
using System.Drawing;
using System.Drawing.Drawing2D;
using System.Windows.Forms;
namespace chp04_10_04
{
    public partial class Form1 : Form
    {
        public Form1()
        {
            InitializeComponent();
        }
        private void button1_Click(object sender, EventArgs e)
        {
            Graphics mygraphic = this.CreateGraphics();
            Pen redPen = new _____①_____ ;         //定义红色画笔
            Pen bluePen = new Pen(Color.Blue,5);
            Pen greenPen = new Pen(Color.Green,2);
            Point p1 = new Point(150, 150);
            Point p2 = new Point(250,150);
            redPen.DashStyle = DashStyle.Dash;
            redPen.Width = 5;
            mygraphic.DrawLine(redPen, 130, 150, 130, 250);
            bluePen.StartCap = LineCap.RoundAnchor;
            _____②_____ = LineCap.ArrowAnchor;   //蓝线尾设置为箭头
            _____③_____ ;                         //从点 p1 至点 p2 绘制蓝线
            mygraphic.DrawLine(greenPen, 160, 160, 250, 250);
        }
    }
}
```

五、编写程序题

1. 设计如图 4-19 所示的图形用户界面程序,单击"绘制棋盘"按钮绘制象棋棋盘。

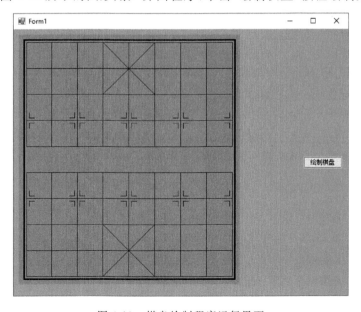

图 4-19 棋盘绘制程序运行界面

2. 编写一个带秒针、分针、时针的"时钟"程序,按计算机时间实时显示,其界面如图 4-20 所示。

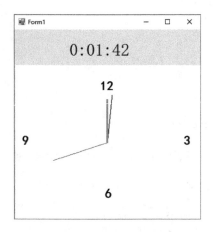

图 4-20 时钟程序运行界面

习题 11 数据库应用

一、选择题

1. 创建数据库连接使用的对象是(   )。
   A. Connection   B. Command   C. DataReader   D. DataSet
2. 若将数据库中的数据填充到数据集,应调用 SqlDataAdapter 的(   )方法。
   A. Open   B. Close   C. Fill   D. Update
3. 若将数据集中所作更改更新回数据库,应调用 SqlAdapter 的(   )方法。
   A. Update   B. Close   C. Fill   D. Open
4. ADO.NET 中的(   )对象主要负责生成并执行 SQL 语句,可以访问用于返回数据、修改数据的数据库命令。
   A. Connection   B. Command   C. DataReader   D. DataSet
5. ADO.NET 中的(   )对象主要负责读取数据库中的数据。
   A. Connection   B. Command   C. DataReader   D. DataSet
6. 在.NET Framework 中可以使用(   )对象连接和访问数据库。
   A. MDI   B. JIT   C. ADO.NET   D. System.ADO
7. 利用 Connection 对象的(   )方法可以打开数据库,建立与数据库的连接。
   A. Close   B. ConnectionString
   C. Database   D. Open
8. 在 ADO.NET 中,以下关于 DataSet 类的叙述错误的是(   )。
   A. 可以向 DataSet 的表集合中添加新表
   B. DataSet 支持 ADO.NET 的不连贯连接及数据分布
   C. DataSet 相当于内存中的一个临时数据库
   D. DataSet 中的数据是只读的

9. 利用 ADO.NET 访问数据库,在联机模式下,不需要使用(　　)对象。
   A. Connection　　B. DataReader　　C. DataAdapter　　D. Command
10. 在脱机模式下,支持离线访问的关键对象是(　　)。
   A. Connection　　　　　　　　B. Command
   C. DataAdapter　　　　　　　 D. DataSet
11. 在 C♯ 程序中,如果需要连接 SQL Server 数据库,则需要使用的连接类是(　　)。
   A. SqlConnection　　　　　　 B. OleDbConnection
   C. OracleConnection　　　　　D. OdbcConnection
12. 利用 Command 对象的 ExecuteNonQuery()方法执行 Insert、Update 或 Delete 语句时,返回(　　)。
   A. true 或 false　　　　　　　B. 1 或 0
   C. 受影响的行数　　　　　　　D. －1
13. 假设已经通过如下 SQL 语句创建表:

```
CREATE TABLE Student
(   id nchar(10) not null,
    Name nchar(10) null,
    Sex nchar(1) null,
    Birthday smalldatetime null,
    Score int null)
```

有以下 C♯ 代码,其功能是读取表中的第二列数据(已知 myreader 为 OleDbDataReader 对象),正确的语句为(　　)。

```
while(myreader.(①))
{  Console.WriteLine(②);}
```

   A. ①Read()　②myreader[1].ToString()
   B. ①Next()　②myreader[1].ToString()
   C. ①Read()　②myreader[2].ToString()
   D. ①Next()　②myreader[2].ToString()

14. 在 ADO.NET 中,执行数据库的某个存储过程,至少需要创建(　　),并设置它们的属性、调用合适的方法。
   A. 一个 Connection 对象和一个 Command 对象
   B. 一个 Command 对象和一个 DataAdapter 对象
   C. 一个 Connection 对象和一个 Dataset 对象
   D. 一个 Command 对象和一个 DataSet 对象

15. ADO.NET 中的 DataView 控件可以用来筛选数据集中的数据项,以下代码用来筛选数据集中分数大于 600 分的学生:

```
DataView mydataview = new DataView(mydataSet.Tables[0]);
(①)
```

则①处应该填写的正确语句为(　　)。
   A. mydataview.RowFilter="Select score from student where score>600";

B. mydataview.RowFilter=”Score＞600”;
C. mydataview.Excute(”Select score from student where score＞600”);
D. mydataview.Excute(”Score＞600”);

二、填空题

1. 数据库通过_____对象连接后,便可以通过_____对象将 SQL 语句(如 SELECT)交由数据库引擎执行,并通过_____对象将数据查询的结果存放到离线的_____对象中,进行离线数据处理。

2. OLEDB.NET 数据提供程序的命名空间是_____。

3. OleDbConnection 类的对象是用来建立与一个数据库的物理连接。建立连接的核心是设置连接字符串_____属性。设置好连接串属性后,可以用该对象的_____方法打开数据库,使用之后,用_____方法关闭与数据库的连接。

4. 使用 OLEDB.NET 访问数据库,已经建立到数据库的连接对象 mycon,并已经打开数据库,从 student 数据表取得所有的记录,需要先创建一个 OleDbCommand 对象 mycmd,其语句是_____。

5. 题 4 中执行查询命令,需要从返回结果集中提取数据,可以使用_____对象,假设将处理的结果放到 myreader 中,其语句为_____。

6. 数据适配器_____对象可以执行 SQL 命令以及调用存储过程、传递参数、获取数据结果集,在数据库和_____数据集对象之间传输数据。

7. 数据适配器中的_____方法用来执行该对象的_____属性中对应的 SQL 语句,将检索出来的数据用来更新数据集 DataSet 中的 DataTable 对象。

8. _____是将数据连接到用户界面控件的过程。在执行该操作后,可以通过窗体控件操作数据库中的数据。

9. _____对象能够创建 DataTable 中所存储数据的不同视图,用于对 DataSet 中的数据进行排序、过滤和查询操作。

10. ADO.NET 中,下列代码运行后,TextBox1.Text 的值是_____。

```
DataTable dt = new DataTable();
dt.Columns.Add("bianhao", typeof(System.Int16));
dt.Columns.Add("chengji", typeof(System.Single));
textBox2.Text = dt.Columns[0].DataType.ToString();
```

三、阅读程序题

1. 阅读下面的程序,说明单击 button1 按钮后程序实现的功能。

```
using System;
using System.Data;
using System.Data.SqlClient;
using System.Windows.Forms;
namespace chp04_11_01
{
    public partial class Form1 : Form
    {
        public Form1()
```

```csharp
{
    InitializeComponent();
}
private void button1_Click(object sender, EventArgs e)
{
    string myconnstr = "Data Source=DESKTOP-331UI15;
                        Initial Catalog=stu;Integrated Security=True";
    string mysql = "insert into student(id,name,sex,birthday,score)
                    values(@id,@name,@sex,@birthday,@score)";
    SqlConnection myconn = new SqlConnection();
    myconn.ConnectionString = myconnstr;
    myconn.Open();
    SqlCommand mycmd = new SqlCommand();
    mycmd.Connection = myconn;
    try
    {
        mycmd.CommandText = mysql;
        mycmd.CommandType = CommandType.Text;
        SqlParameter p1 = mycmd.CreateParameter();
        p1.ParameterName = "@id"; p1.Value = "0010";
        mycmd.Parameters.Add(p1);
        SqlParameter p2 = mycmd.CreateParameter();
        p2.ParameterName = "@name"; p2.Value = "刘新";
        mycmd.Parameters.Add(p2);
        SqlParameter p3 = mycmd.CreateParameter();
        p3.ParameterName = "@sex"; p3.Value = "男";
        mycmd.Parameters.Add(p3);
        SqlParameter p4 = mycmd.CreateParameter();
        p4.ParameterName = "@birthday"; p4.Value = "1997-09-09";
        mycmd.Parameters.Add(p4);
        SqlParameter p5 = mycmd.CreateParameter();
        p5.ParameterName = "@score"; p5.Value = "611";
        mycmd.Parameters.Add(p5);
        mycmd.ExecuteNonQuery();
    }
    catch(Exception ex)
    {
        MessageBox.Show(ex.Message);
    }
}
```

2. 下面的程序对数据库 stu 中的 student(id,name,sex,birthday,score) 表进行操作，其运行界面如图 4-21 所示，请写出单击 button1 按钮后文本框的内容。

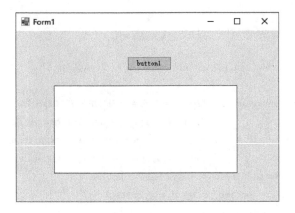

图 4-21 数据表操作程序运行界面

```
using System;
using System.Data.SqlClient;
using System.Windows.Forms;
namespace chp04_11_02
{
    public partial class Form1 : Form
    {
        public Form1()
        {
            InitializeComponent();
        }
        private void button1_Click(object sender, EventArgs e)
        {
            SqlConnection myconn = new SqlConnection();
            SqlCommand mycmd = new SqlCommand();
            string strconn = "Data Source = DESKTOP - 331UI15;
                    Initial Catalog = stu; Integrated Security = True";
            textBox1.Text = "";
            myconn.ConnectionString = strconn;
            myconn.Open();
            mycmd.Connection = myconn;
            mycmd.CommandText = "select * from student";
            SqlDataReader result = mycmd.ExecuteReader();
            for (int i = 0; i < result.FieldCount; i++)
                textBox1.Text += result.GetName(i) + "  ";
            result.Close();
            myconn.Close();
        }
    }
}
```

四、程序填空题

1. 以下程序运行后单击"装入数据"按钮,将数据表 student 的内容导入到

DataGridView 中。请完善程序(程序运行界面见图 4-22)。

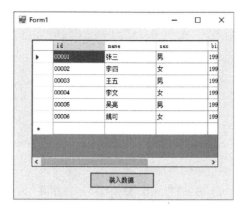

图 4-22　DataGridView 应用程序运行界面

```
using System;
using System.Data;
using System.Data.OleDb;
using System.Windows.Forms;
namespace chp04_11_03
{
    public partial class Form1 : Form
    {
        public Form1()
        {
            InitializeComponent();
        }
        private void button1_Click(object sender, EventArgs e)
        {
            OleDbConnection mydbconn = new OleDbConnection("Provider =
                        SQLNCLI11;Data Source = DESKTOP-331UI15;
                        Integrated Security = SSPI;Initial Catalog = stu");
            try
            {
                OleDbDataAdapter myda = new
                        OleDbDataAdapter("select * from student", mydbconn);
                DataTable mydt = new DataTable("student");
                myda.Fill(    ①    );
                this.dataGridView1.DataSource =      ②     ;
            }
            catch(Exception ey)
            {
                MessageBox.Show(ey.Message);
            }
        }
    }
}
```

2. 设计一个模拟用户登录的程序,界面如图 4-23 所示,用户数据存储数据库 stu 中的表 st_user 中。请完善程序。

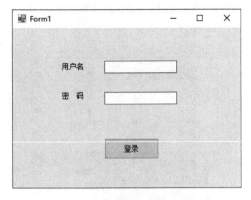

图 4-23 用户登录程序运行界面

```
using System;
using System.Data.SqlClient;
using System.Windows.Forms;
namespace chp04_11_04
{
    public partial class Form1 : Form
    {
        public Form1()
        {
            InitializeComponent();
        }
        private void button1_Click(object sender, EventArgs e)
        {
            string strconn = "Data Source = DESKTOP - 331UI15;
                Initial Catalog = stu;Integrated Security = True";
            SqlConnection myconn = _____①_____;
            myconn.ConnectionString = strconn;
            _____②_____;
            string mysql = String.Format("select count ( * ) from st_user where
            user_name = '{0}' and user_pass = '{1}'", textBox1.Text, textBox2.Text);
            SqlCommand mycomm = new SqlCommand();
            mycomm.Connection = myconn;
            mycomm.CommandText = _____③_____;
            int num = (int)mycomm.ExecuteScalar();
            if (num > 0)
                MessageBox.Show("身份验证通过","提示");
            else
                MessageBox.Show("身份验证没通过","提示");
        }
    }
}
```

五、编写程序题

1. 设计一个实现模糊查询的程序,可以按照学号或者姓名或者分数查询 Student 表中的内容,Student 表结构为 student(id,name,sex,birthday,score),程序运行界面如图 4-24 所示。

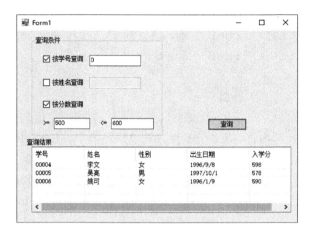

图 4-24　模糊查询程序运行界面

2. 使用 Command 对象访问 SQL-Server 数据库的 student(id,name,sex,birthday,score)表,设计一个界面如图 4-25 所示的应用程序并完成相应代码。

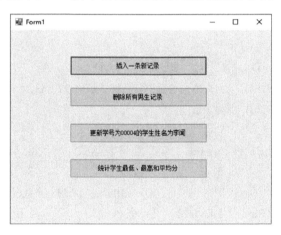

图 4-25　数据表操作程序运行界面

参 考 答 案

习题 1

一、选择题

1. D　　2. B　　3. B　　4. A　　5. A　　6. C　　7. D　　8. D　　9. A　　10. A

二、填空题

1. MSIL　　　　2. Main　　　　3. 公共语言运行库(CLR)

4. 面向对象

习题 2

一、选择题

1. C　　2. C　　3. A　　4. A　　5. D　　6. B　　7. B　　8. D　　9. C　　10. B

11. D 12. B 13. C 14. A 15. A 16. C 17. A 18. C 19. B 20. D
21. A 22. B 23. D 24. C 25. D 26. C 27. A 28. A 29. D 30. C
31. A 32. C 33. A 34. B 35. B

二、填空题

1. &&   2. +=
3. //   4. true false   5. 变量,自右至左
6. 第一个输出 RP 后换行,第二个不换   7. Main   8. float double decimal
9. ReadLine   10. 3   11. －123
12. True   13. 1   14. double
15. Math.Pow(Math.Sin(x),2)＊(x＋y)/(x－y) 或 Math.Sin(x)＊Math.Sin(x)＊(x＋y)/(x－y)
16. 单引号,一,双引号,字符
17. true   18. byte   19. 键盘,显示器
20. 2.1   21. true true   22. 10,6
23. －2   24. －8

习题 3

一、选择题

1. B 2. B 3. D 4. A 5. C 6. D 7. A 8. B 9. C 10. D
11. A 12. B 13. C 14. A 15. C 16. D 17. C 18. D 19. B 20. C
21. D 22. D 23. B 24. B

二、填空题

1. for、while、do while 和 foreach   2. 常量表达式   3. break
4. default   5. 2   6. 7   7. false   8. x<z‖y<z
9. x<－10‖x>2&&x<3   10. (ch>='A')&&(ch<='Z')
11. false   12. 135   13. 642   14. 条件   15. 三,从右至左
16. min=(a<b)? a:b;   17. true   18. 5,6   19. 顺序结构、选择结构、循环结构
20. Continue   21. 13,30   22. 2、1、3   23. m=14,m=11,m=8

三、阅读程序题

1. m=5   2. k=1   3. 1,4;2,6;5,6;c=24   4. ＋＊++p=20

四、程序填空题

1. ① f1+f2   ② i%5==0   ③ f2
2. ① s=0   ② n>500   ③ n%11==0
3. ① n＊n%100   ② else   ③ sqr==n
4. ① b++   ② 20－a－b   ③ 5＊a+4＊b+2＊c==50
5. ① sum+a   ② q   ③ sum<100
6. ① break   ② d＊n   ③ rmb1－sum

五、编写程序题

1. 参考程序如下:

using System;

```
namespace chp04_03_11
{
    class Program
    {
        static void Main(string[] args)
        {
            int i, n;
            long fact = 1;
            Console.WriteLine("请输入一个正整数:");
            n = int.Parse(Console.ReadLine());
            for (i = 2; i <= n; i++)
                fact *= i;
            Console.WriteLine("{0}!={1}", n, fact);

        }
    }
}
```

注意,当 n 较大时,上面程序会得到错误结果,但不会给出提示信息。如下程序在不能求 n!时会输出提示信息。

```
using System;
namespace chp04_03_12
{
    class Program
    {
        static void Main(string[] args)
        {
            int i, n, over = 0;
            long  fac = 1, fac2 = 1;
            Console.WriteLine("请输入一个正整数:");
            n = int.Parse(Console.ReadLine());
            for (i = 1; i <= n; i++)
            {
                fac2 = fac * i;
                if (fac != fac2 / i)        //判断值是否超过变量能表示的最大数
                {
                    over = 1; break;
                }
                fac = fac2;
                Console.WriteLine("{0}!={1}", i, fac);   //此语句可监视 i! 阶乘
            }
            if (over == 0)
                Console.WriteLine("{0}!={1}", n, fac);
            else
                Console.WriteLine("太大不能求阶乘");
        }
    }
}
```

当 n 较大时,可近似求 n!。

当计算出的阶乘值大于等于 10 时,就除以 10,然后指数加 1。最后将尾数和指数分别输出。例如,计算 1000000! 的程序如下:

```csharp
using System;
namespace chp04_03_13
{
    class Program
    {
        static void Main(string[] args)
        {
            double fact = 1.0;                    //存储阶乘的变量 fact 声明为 double 型,
                                                  //可以得到更多的有效位数
            int e = 0;
            for (int i = 2; i <= 1000000; i++)
            {
                fact *= i;
                while (fact >= 10)                //阶乘反复除以 10
                {
                    fact /= 10.0;
                    e++;
                }
            }
            Console.WriteLine("{0}e{1}",fact,e);  //输出 8.26393e5565708
        }
    }
}
```

当 n 较大时,要求 n! 精确值,要寻求其他方法。

2. 参考程序如下:

```csharp
using System;
namespace chp04_03_14
{
    class Program
    {
        static void Main(string[] args)
        {
            int i, j, k, count = 0;
            for (i = 9; i >= 1; i--)
                for (j = 9; j >= 0; j--)
                    if (i == j) continue;
                    else
                        for (k = 0; k <= 9; k++)
                            if ((k != i) && (k != j)) count++;
            Console.WriteLine("{0}",count);
        }
    }
}
```

3. 分析:球只能是五种颜色之一。设取出的球为 i、j、k,根据题意,i、j、k 分别可以有五种取值,且 i≠j≠k。采用穷举法,逐个验证每一种可能的组合,从中找出符合要求的组合并输出。

参考程序如下：

```
using System;
namespace chp04_03_15
{
    class Program
    {
        static void Main(string[ ] args)
        {
            const int red = 0, yellow = 1, blue = 2, white = 3, black = 4;    //五种颜色
            int print = 0;                           //打印球的颜色
            int n, loop, i, j, k;
            n = 0;
            for (i = red; i <= black; i++)
            {
                for (j = red; j <= black; j++)
                {
                    if (i != j)
                    {
                        for (k = red; k <= black; k++)
                        {
                            if ((k != i) && (k != j))
                            {
                                n = n + 1;
                                Console.Write("{0,4}",n);
                                for (loop = 1; loop <= 3; loop++)
                                {
                                    switch (loop)
                                    {
                                      case 1: print = i; break;
                                      case 2: print = j; break;
                                      case 3: print = k; break;
                                        default: break;
                                    }
                                    switch (print)
                                    {
                                        case red: Console.Write("      red"); break;
                                        case yellow: Console.Write("   yellow"); break;
                                        case blue: Console.Write("    blue"); break;
                                        case white: Console.Write("   white"); break;
                                        case black: Console.Write("   black"); break;
                                        default: break;
                                    }
                                }
                                Console.WriteLine();
                            }
                        }
                    }
                }
            }
        }
```

```
            Console.WriteLine("total:{0}",n);
        }
    }
}
```

4．参考程序如下：

```
using System;
namespace chp04_03_16
{
    class Program
    {
        static void Main(string[ ] args)
        {
            int row;                                    //菱形行数
            Console.WriteLine("请输入行数：");
            row = int.Parse(Console.ReadLine());
            int i, j, n;
            n = row / 2 + 1;
            for (i = 1; i <= n; i++)                    //输出前 n 行图案
            {
                for (j = 1; j <= n - i; j++)            //输出 * 字符前面的空格
                    Console.Write(" ");
                for (j = 1; j <= 2 * i - 1; j++)        //循环输出字符 *
                    Console.Write(" * ");
                Console.WriteLine();
            }
            for (i = 1; i <= n - 1; i++)                //输出后 row-n 图案
            {
                for (j = 1; j <= i; j++)
                    Console.Write(" ");
                for (j = 1; j <= row - 2 * i; j++)
                    Console.Write(" * ");
                Console.WriteLine();
            }
        }
    }
}
```

习题 4

一、选择题

1．A　　2．B　　3．B　　4．C　　5．D　　6．B　　7．C　　8．D　　9．B　　10．A
11．A　　12．D　　13．C　　14．A　　15．A

二、填空题

1．继承　　2．private　　3．构造函数　　4．封装　　5．析构函数　　6．类
7．value　　8．return　　9．Get　　10．方法　　11．static　　12．同名
13．new　　14．构造函数

三、阅读程序题

1．i=1, k = 1
　 i=1, k = 2

2. 10   10   10   10
   9   8   7   6

3. 0,10,20
   a=0,b=10,c=20

4. 024

5. 050055

6. 5
   11
   0
   str
   Hello,World!

7. 0
   13

四、程序填空题

1. ① new A1()   ② a.display()
2. ① p.N   ② return p_name;   ③ p_name=value

五、编写程序题

1. 参考程序如下：

```
using System;
namespace chp04_04_10
{
    class rectangle
    {
        int length, width;
        public  rectangle(int i = 0, int j = 0)
        { length = i; width = j; }
        public int Length
        {
            get { return length; }
            set
            {
                if (value > 0)
                    length = value;
                else
                    length = 0;
            }
        }
        public int Width
        {
            get
            {
                return width;
            }
            set
            {
```

```csharp
                if (value > 0)
                    width = value;
                else
                    width = 0;
            }
        }
         public int area()
        {
            int s = length * width;
            return s;
        }
        public int fun()
        {
            int p = 2 * (length + width);
            return p;
        }
    };
    class Program
    {
        static void Main(string[] args)
        {
            int x, y;
            Console.WriteLine("请输入矩形的长与宽");
            x = int.Parse(Console.ReadLine());
            y = int.Parse(Console.ReadLine());
            rectangle a = new rectangle(x, y);
            Console.WriteLine("周长和面积为:{0}\t{1}", a.area(), a.fun());
            Console.WriteLine("请输入矩形的长与宽");
            x = int.Parse(Console.ReadLine());
            y = int.Parse(Console.ReadLine());
            a.Length = x;
            a.Width = y;
            Console.WriteLine("周长和面积为:{0}\t{1}", a.area(), a.fun());
        }
    }
}
```

2. 参考程序如下:

```csharp
using System;
namespace chp04_04_11
{
    class employee
    {
        protected  string name;              //姓名
        protected string section;            //部门
        protected string title;              //职称
        public  employee(string _name, string _selection, string _title)
        {
            name = _name;
            section = _selection;
```

```
                title = _title;
            }
            public void changename(string str)
            {
                name = str;
            }
            public void display()
            {
                Console.WriteLine("{0}\t{1}\t{2}", name,section,title);
            }
        };
        class Program
        {
            static void Main(string[] args)
            {
                employee obj1 = new employee("李明", "计算机系", "高级工程师");
                employee obj2 = new employee("张闻毅", "计算中心", "副教授");
                employee obj3 = new employee("刘益录", "计算机系", "教授");
                obj1.display();
                obj2.display();
                obj3.display();
            }
        }
    }
```

3. 参考程序如下：

```
using System;
namespace chp04_04_12
{
    class toy
    {
        string name;
        float Price, Total;
        int Count;
        public toy() { }
        public toy(string n, float p, int c)
        {
            name = n;
            Price = p;
            Count = c;
        }
        public void input(float P, int C)
        {
            Price = P;
            Count = C;
        }
        public void compute()
        {
            Total = (float)Price * Count;
        }
```

```csharp
        public void display()
        {
            Console.WriteLine("{0}:Price = {1}, Count = {2}, Total = {3}", name, Price, Count, Total);
        }
    };
    class Program
    {
        static void Main(string[] args)
        {
            toy a1 = new toy("Car",21.5f, 12);
            toy a2 = new toy("Bicycle",12.8f, 20);
            a1.compute();
            a1.display();
            a2.compute();
            a2.display();
        }
    }
}
```

4. 参考程序如下：

```csharp
using System;
namespace chp04_04_13
{
    class Myclock
    {
        int hours, minutes, seconds;
        public Myclock(int h, int m, int s)
        {
            hours = h;
            minutes = m;
            seconds = s;
        }
        public void Addhours()
        {
            hours = ++hours % 24;
        }
        public void Addminutes()
        {
            ++minutes;
            if (minutes >= 60)
            {
                minutes = 0;
                Addhours();
            }
        }
        public void Addseconds()
        {
            ++seconds;
            if (seconds >= 60)
            {
```

```
                seconds = 0;
                Addminutes();
            }
        }
        public void display()
        {
            Console.WriteLine("Hours = {0},Minutes = {1},Seconds = {2}", hours, minutes, seconds);
        }
    }
    class Program
    {
        static void Main(string[] args)
        {
            Myclock c1 = new Myclock(15, 59, 59);
            c1.display();
            Console.WriteLine("add 1 seconds:");
            c1.Addseconds();
            c1.display();
            Myclock c2 = new Myclock(23, 59, 30);
            c2.display();
            Console.WriteLine("add 1 minutes");
            c2.Addminutes();
            c2.display();
        }
    }
}
```

5. 参考程序如下：

```
using System;
namespace chp04_04_14
{
    class Triangle
    {
        int x1, x2, x3, y1, y2, y3;
        public Triangle(int mx1, int my1, int mx2, int my2, int mx3, int my3)
        {
            x1 = mx1;
            y1 = my1;
            x2 = mx2;
            y2 = my2;
            x3 = mx3;
            y3 = my3;
        }
        public bool IsTriangle()
        {
            if ((y3 - y2) * (x3 - x1) == (y3 - y1) * (x3 - x2))
                return false;
            else
                return true;
        }
```

```csharp
            public double Area()
            {
                double area;
                area = Math.Abs(x1 * y2 + x2 * y3 + x3 * y1 - y1 * x2 - y2 * x3 - y3 * x1) / 2.0;
                return area;
            }
        }
        class Program
        {
            static void Main(string[] args)
            {
                Triangle t1 = new Triangle(10, 5, 20, 10, 15, 1);
                Triangle t2 = new Triangle(10, 5, 20, 5, 15, 5);
                if (t1.IsTriangle())
                    Console.WriteLine("t1 三角形面积 = {0}", t1.Area());
                else
                    Console.WriteLine("t1 这三点不能构成三角形");
                if (t2.IsTriangle())
                    Console.WriteLine("t2 该三角形面积 = {0}", t1.Area());
                else
                    Console.WriteLine("t2 这三点不能构成三角形");
            }
        }
    }
```

习题 5

一、选择题

1. D　2. B　3. B　4. D　5. C　6. C　7. A　8. C　9. A　10. D
11. B　12. A　13. B　14. D　15. C　16. B　17. B　18. A　19. B　20. A
21. C　22. A　23. D　24. D　25. A

二、填空题

1. 当前类　　　2. 多态　　　3. abstract　　4. 传递性　　5. 字段,属性,方法
6. 默认构造函数　7. 接口　　　8. Delegate　　9. virtual　　10. 实现接口的类

三、阅读程序题

1. eat cheese! mouse sleeping!

2. begin create B object
   begin create A object
   begin destory A object
   begin destory B object

3. 非默认构造函数
   abc
   基类默认构造函数!
   派生类构造函数!

4. 静态构造函数!
   构造函数!

构造函数！

5. Cat cryed!

　　mouse1 attempt to escape!

　　mouse2 attempt to escape!

　　Host waken!

四、程序填空题

1. ① new Triangle(3,4)　　② t.area()　　③ Shape
2. ① A a1＝new A(x)　　② base(x)　　③ c＝x＋2
3. ① i++　　② ＜　　　　　　　　4. ① return －1　　② table
5. ① s[i]　　② '\0'或　0　　　　　6. ① 2　　② 2

五、编写程序题

1. 参考程序如下：

```
using System;
namespace chp04_05_08
{
    class Building
    {
        public Building(int f, int r, double ft)
        {
            floors = f;
            rooms = r;
            footage = ft;
        }
        public void show()
        {
            Console.WriteLine(" floors:{0}",floors);
            Console.WriteLine(" rooms:{0}",rooms);
            Console.WriteLine(" total area:{0}",footage);
        }
        protected int floors;
        protected int rooms;
        protected double footage;
    };
    class Housing :Building
    {
        public Housing(int f, int r, double ft, int bd, int bth):base(f, r, ft)
        {
            bedrooms = bd;
            bathrooms = bth;
        }
        public void show()
        {
            Console.WriteLine("HOUSING:");
            base.show();
            Console.WriteLine(" bedrooms:{0}",bedrooms);
            Console.WriteLine(" bathrooms:{0}", bathrooms);
        }
```

```csharp
            privateint bedrooms;
            private int bathrooms;
        };
        class Office : Building
        {
        public Office(int f, int r, double ft, int ph, int ex):base(f, r, ft)
        {
            phones = ph;
            extinguishers = ex;
        }
        public void show()
        {
            Console.WriteLine("HOUSING:");
            base.show();
            Console.WriteLine(" phones:{0}",phones);
            Console.WriteLine(" extinguishers:{0}",extinguishers);
        }
        privateint phones;
        private int extinguishers;
        };
        class Program
            {
                static void Main(string[] args)
                {
                    Housing hob = new Housing(5, 7, 140, 2, 2);
                    Office oob = new Office(8, 12, 500, 12, 2);
                    hob.show();
                    oob.show();
                }
            }
        }
```

2. 参考程序如下：

```csharp
using System;
namespace chp04_05_09
{
    class Shape
    {
        protected double s;              //面积
        public void show()
        {
            Console.WriteLine("s = {0}",s);
        }
    };
    class Square :Shape
    {
        protected double a;
        public Square(double a1 = 0) { a = a1; }
        void SetArea() { s = a * a; }
    };
```

```
class Circle : Shape
{
    double r;
    public Circle(double r1 = 0) { r = r1; }
    public void SetArea() { s = r * r * 3.14; }
};
class Rectangle : Square
{
    double b;                      // 增加另一成员表示边长
    public Rectangle(double x1 = 0, double y1 = 0) : base(x1) { b = y1; }
    public void SetArea() { s = a * b; }
};
class Program
{
    static void Main(string[] args)
    {
        Rectangle r = new Rectangle(3, 4);
        r.SetArea();
        r.show();
    }
}
```

3. 类(含部分成员)定义如下：

```
class person                          //定义一个基类
{
    protected string name;            //姓名
    protected string sex;             //性别
    protected int age;                //年龄
        …
};
class teacher :person                 //基类派生出教师类
{
    int sno;                          //工号
    string zc;                        //职称
    double wages;                     //工资
        …
};
class student :person                 //基类派生出学生类
{
    int sno;                          //学号
    string bj;                        //班级
    string zy;                        //专业
    double score;                     //入学成绩
        …
};
```

习题 6

一、选择题

1. C   2. B   3. B   4. C   5. D   6. D   7. B   8. A   9. B   10. C
11. D   12. D   13. A   14. C   15. B   16. D

二、填空题

1. Length   2. foreach   3. 引用   4. 交错数组   5. 0
6. null   7. System.Enum   8. 4
9. float[,] f=new float[3,4];   10. @   11. Sort   12. 246

三、阅读程序题

1. the sum is 4
2. Cat cryed!
   mouse1 attempt to escape!
   mouse2 attempt to escape!
   Host waken!
3. 01234567
4. 9 12 15
5. 115 150 198 30 28
6. Microsoft Studio Visual

四、程序填空题

1. ① t[k] == val   ② i = −1   ③ ref index
2. ① 3−i   ② a[j+1]
3. ① a[i] > m   ② a[i]<n
4. ① n=0   ② sum += a[i];   ③ a[i] > avg

五、编写程序题

1. 参考程序如下：

```
using System;
namespace chp04_06_11
{
    class Program
    {
        static void Main(string[] args)
        {
            int[,] a = new int[4, 5] { { 1, 2, 3, 4, 5 }, { 6, 7, 8, 9, 0 }, { 11, 12, 13, 14, 15
                                     }, { 16, 17, 18, 19, 20 } };
            int m = 0, n = 0;
            for (int i = 0; i < 4; i++)
                for (int j = 0; j < 5; j++)
                    if (a[i, j] % 2 == 0)
                        m++;
                    else
                        n++;
            Console.WriteLine("奇数个数是{0},偶数个数是{1}", n, m);
        }
    }
}
```

2. 参考程序如下：

```csharp
using System;
namespace chp04_06_12
{
    class Program
    {
        static void Main(string[] args)
        {
            int[,] a = new int[3, 3] { { 1, 2, 3 }, { 4, 5, 6 }, { 7, 8, 9 } };
            int sum = 0;
            for (int i = 0; i < 3; i++)
                for (int j = 0; j < 3; j++)
                    if (i == j)
                        sum += a[i, j];
            Console.WriteLine(sum);
        }
    }
}
```

3. 参考程序如下：

```csharp
using System;
namespace chp04_06_13
{
    class Program
    {
        enum colors { red, blue, green };
        static void show(colors color)
        {
            switch (color)
            {
                case colors.red: Console.Write("red"); break;
                case colors.blue: Console.Write("blue"); break;
                case colors.green: Console.Write("green"); break;
            }
            Console.Write('\t');
        }
        static void Main(string[] args)
        {
            colors col1, col2, col3;
            for (col1 = colors.red; col1 <= colors.green; col1 = col1 + 1)
                for (col2 = colors.red; col2 <= colors.green; col2 = col2 + 1)
                    for (col3 = colors.red; col3 <= colors.green; col3 = col3 + 1)
                    {
                        show(col1);
                        show(col2);
                        show(col3);
                        Console.WriteLine();
                    }
        }
    }
}
```

}

4. 分析：关键在于是否为闰年。

对于非闰年，则天数＝1月天数＋2月天数＋……＋上月天数＋日数。

对于闰年，有两种情况

(1) 若月份≥3，则天数＝1月天数＋2月天数＋……＋上月天数＋日数＋1；

(2) 若月份＜3，则天数与非闰年一致。

参考程序如下：

```
using System;
namespace chp04_06_14
{
    class Program
    {
        struct MyDate
        {
            public int month, day, year;
        }
        static void Main(string[] args)
        {
            int i, days;
            MyDate mydate;
            int[] daytab = new int[] { 0, 31, 28, 31, 30, 31, 30, 31, 31, 30, 31, 30, 31 };
            Console.Write("Enter day,month,year:");
            mydate.day = int.Parse(Console.ReadLine());
            mydate.month = int.Parse(Console.ReadLine());
            mydate.year = int.Parse(Console.ReadLine());
            days = 0;
            for(i = 1; i < mydate.month; i++)
                days += daytab[i];
            days += mydate.day;
            if(((mydate.year % 4 == 0)&&(mydate.year % 100!= 0) ||
                (mydate.year % 400 == 0))&&(mydate.month >= 3))
                days += 1;
            Console.WriteLine("month {0},day {1},is the {2}th day in year
                    {3}",mydate.month,mydate.day,days,mydate.year);
        }
    }
}
```

5. 分析：定义一个候选人结构体数组，包括3个元素，在每个元素中存放有关的数据。

参考程序如下：

```
using System;
namespace chp04_06_15
{
    struct candidate
    {
        public string name;
        public int count;
    }
    class Program
    {
```

```
        static void Main(string[] args)
        {
            candidate[] ca = new candidate[3];
            String[] names = new String[] { "Li", "Zhang", "Fun" };
            int i ,j;
            for ( i = 0; i <= 2; i++)
            { ca[i].name = names[i];
                ca[i].count = 0; }
            string votename;        //votename是投票人所选的人的姓名
            for (i = 0; i < 10; i++)
            {
                votename = Console.ReadLine();
                for (j = 0; j < 3; j++)
                    if (votename == ca[j].name) ca[j].count++;
            }
            for (i = 0; i < 3; i++)
                Console.WriteLine("{0}:{1}",ca[i].name,ca[i].count);
        }
    }
}
```

习题 7

一、选择题

1. C   2. B   3. B   4. D   5. A   6. C   7. B   8. D   9. B   10. D
11. A   12. C   13. A   14. A   15. B   16. B   17. A   18. C   19. C   20. C

二、填空题

1. Interval   2. ListBox1.Items.Count   3. 0，−1
4. Text   5. Clicked   6. PasswordChar
7. 视图   8. Enabled，false   9. visible，false
10. Image，SizeMode，StretchImage   11. Value   12. RichTextBox
13. ComboBox1.Items.Add("长沙")   14. SelectionMode   15. DateSelected

三、阅读程序题

1. 单击 button1 按钮后的程序运行结果如图 4-26 所示。

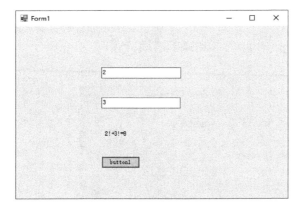

图 4-26　文本框应用程序运行界面

2. 单击 button1 按钮后的程序运行结果如图 4-27 所示。

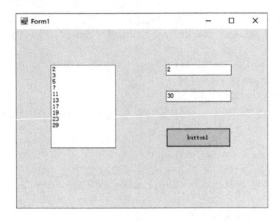

图 4-27　列表框应用程序运行界面

3. 单击 button1 按钮后的程序运行结果如图 4-28 所示。

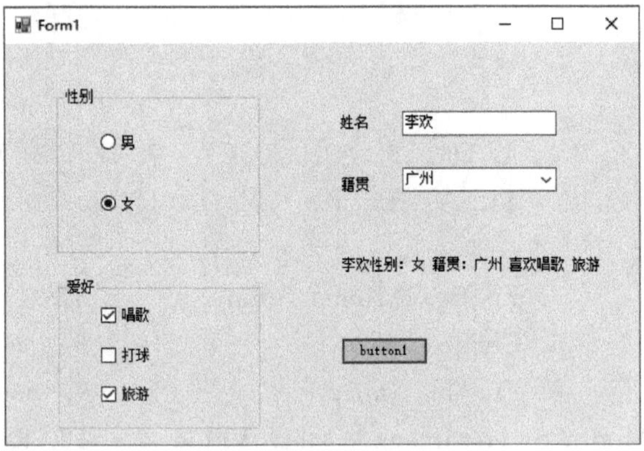

图 4-28　单选按钮和复选框应用程序运行界面

4. 单击 button1 按钮后的程序运行结果如图 4-29 所示。

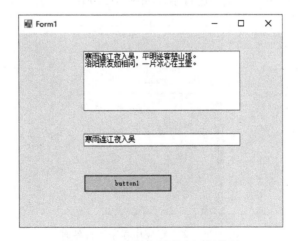

图 4-29　文本框应用程序运行界面

四、程序填空题

1. ① s = s * 10 + j % 10   ② huiwen(i) == true   ③ i.ToString()
2. ① i<listBox1.Items.Count； ② j<=a.Length-1
   ③ listBox2.Items.Add(a[i].ToString())

五、编写程序题

1. 参考程序如下：

```
using System;
using System.Windows.Forms;
namespace chp04_07_07
{
    public partial class Form1 : Form
    {
        private string strop;
        private double num1, num2;
        private Button btn;
        public Form1()
        {
            InitializeComponent();
        }
        private void Form1_Load(object sender, EventArgs e)
        {
            textBox1.Text = "";
            label1.Text = "";
        }
        private void process()
        {
            num1 = double.Parse(textBox1.Text);
            label1.Text = textBox1.Text + strop;
            textBox1.Text = "";
        }
        private void button2_Click(object sender, EventArgs e)
        {
            textBox1.Text = textBox1.Text + button2.Text;
        }
        private void button3_Click(object sender, EventArgs e)
        {
            textBox1.Text = textBox1.Text + button3.Text;
        }
        private void button4_Click(object sender, EventArgs e)
        {
            textBox1.Text = textBox1.Text + button4.Text;
        }
        private void button5_Click(object sender, EventArgs e)
        {
            textBox1.Text = textBox1.Text + button5.Text;
        }
        private void button6_Click(object sender, EventArgs e)
        {
```

```csharp
            textBox1.Text = textBox1.Text + button6.Text;
        }
        private void button7_Click(object sender, EventArgs e)
        {
            textBox1.Text = textBox1.Text + button7.Text;
        }
        private void button8_Click(object sender, EventArgs e)
        {
            textBox1.Text = textBox1.Text + button8.Text;
        }
        private void button9_Click(object sender, EventArgs e)
        {
            textBox1.Text = textBox1.Text + button9.Text;
        }
        private void button10_Click(object sender, EventArgs e)
        {
            textBox1.Text = textBox1.Text + button10.Text;
        }
        private void button11_Click(object sender, EventArgs e)
        {
            textBox1.Text = textBox1.Text + button11.Text;
        }
        private void button12_Click(object sender, EventArgs e)
        {
            if (label1.Text == "")
                MessageBox.Show("输入不正确", "信息提示",
                    MessageBoxButtons.OK);
            else
            {
                num2 = double.Parse(textBox1.Text);
                label1.Text = label1.Text + textBox1.Text;
                switch(strop)
                {
                    case "+":
                        textBox1.Text = (num1 + num2).ToString();
                        break;
                    case "-":
                        textBox1.Text = (num1 - num2).ToString();
                        break;
                    case "*":
                        textBox1.Text = (num1 * num2).ToString();
                        break;
                    case "/":
                        textBox1.Text = (num1 / num2).ToString();
                        break;
                }
                label1.Text = label1.Text + " = " + textBox1.Text;
            }
        }
        private void button13_Click(object sender, EventArgs e)
        {
```

```
            strop = "+";
            process();
        }
        private void button14_Click(object sender, EventArgs e)
        {
            strop = "-";
            process();
        }
        private void button15_Click(object sender, EventArgs e)
        {
            strop = "*";
            process();
        }
        private void button16_Click(object sender, EventArgs e)
        {
            strop = "/";
            process();
        }
        private void button17_Click(object sender, EventArgs e)
        {
            textBox1.Text = "";
            label1.Text = "";
        }
        private void button1_Click(object sender, EventArgs e)
        {
            textBox1.Text = textBox1.Text + button1.Text;
        }
    }
}
```

2. 参考程序如下：

```
using System;
using System.Windows.Forms;
namespace chp04_07_08
{
    public partial class Form1 : Form
    {
        public Form1()
        {
            InitializeComponent();
        }
        private void button2_Click(object sender, EventArgs e)
        {
            richTextBox1.Text += "姓名：" + textBox1.Text + "\n";
            richTextBox1.Text += "性别：" + comboBox1.Text + "\n";
            richTextBox1.Text += "出生日期:" + dateTimePicker1.Text + "\n";
            richTextBox1.Text += "学位：";
            if (radioButton1.Checked == true)
                richTextBox1.Text += radioButton1.Text;
            else if (radioButton2.Checked == true)
```

```
                richTextBox1.Text += radioButton2.Text;
            else if (radioButton3.Checked == true)
                richTextBox1.Text += radioButton3.Text;
            richTextBox1.Text += "\n爱好：";
            if (checkBox1.Checked == true)
                richTextBox1.Text += checkBox1.Text + " ";
            if (checkBox2.Checked == true)
                richTextBox1.Text += checkBox2.Text + " ";
            if (checkBox3.Checked == true)
                richTextBox1.Text += checkBox3.Text;
            richTextBox1.Text += "\n" + "专业：" + listBox1.Text;
        }
        private void button1_Click(object sender, EventArgs e)
        {
            textBox1.Text = "";
            richTextBox1.Text = "";
            radioButton1.Checked = false;
            radioButton2.Checked = false;
            radioButton3.Checked = false;
            checkBox1.Checked = false;
            checkBox2.Checked = false;
            checkBox3.Checked = false;
        }
    }
}
```

习题 8

一、选择题

1. C　2. B　3. C　4. A　5. D　6. B　7. C　8. B　9. D　10. B

二、填空题

1. ContextMenuStrip，ContextMenuStrip，contextMenu1

2. MenuStrip，MenuStrip，ToolStripMenuItem　　3. Checked，true，false

4. OpenFileDialog　　　　　　　　　　　　　　5. FontDialog，ShowDialog

6. SaveFileDialog，FileName

三、阅读程序题

（略）

四、程序填空题

1. ① "*.txt|*.txt"　　　　　　　　　② DialogResult.OK
   ③ saveFileDialog1.FileName　　　　④ myfile.Close()

2. ① DialogResult.OK　　　　　　　　② this.fontDialog1.Font
   ③ DialogResult.OK　　　　　　　　④ colorDialog1.Color

五、编写程序题

1. 参考程序如下：

```
using System;
using System.Windows.Forms;
```

```csharp
namespace chp04_08_05
{
    public struct st
    {
        public string id;
        public string name;
        public string sex;
        public string myclass;
        public int age;
        public st(string sid, string sname, string ssex, string smyclass, int sage)
        {
            id = sid;
            name = sname;
            sex = ssex;
            myclass = smyclass;
            age = sage;
        }
    }
    public partial class Form1 : Form
    {
        public int pos = 0;
        st[] student = {new st("01","李磊","M","1",20 ),
                        new st ("02", "刘浏", "M", "2", 19),
                        new st("03", "张豪", "M", "1", 20),
                        new st( "04", "王佳佳", "F", "2", 21 ),
                        new st( "05", "黄嫣然", "F", "1", 20 ),
                        new st( "06", "李君", "F", "1", 20 ),
                        new st( "07", "马冬", "M", "2", 18 ),
                        new st( "08", "杨柳", "F", "2", 20 )};
        public Form1()
        {
            InitializeComponent();
        }
        public void setvalue(int pos)
        {
            textBox1.Text = student[pos].id;
            textBox2.Text = student[pos].name;
            textBox3.Text = student[pos].sex;
            textBox4.Text = student[pos].myclass;
            textBox5.Text = student[pos].age.ToString();
        }
        private void Form1_Load(object sender, EventArgs e)
        {
            setvalue(0);
        }
        private void toolStripButton1_Click(object sender, EventArgs e)
        {   pos = 0;
            setvalue(pos);
```

```csharp
        }
        private void toolStripButton2_Click(object sender, EventArgs e)
        {
            pos++;
            if (pos <= student.Length - 1)
                setvalue(pos);
            else
                pos--;
        }
        private void toolStripButton3_Click(object sender, EventArgs e)
        {
            pos--;
            if (pos >= 0)
                setvalue(pos);
            else
                pos++;
        }
        private void toolStripButton4_Click(object sender, EventArgs e)
        {
            pos = student.Length - 1;
            setvalue(pos);
        }
    }
}
```

2. 参考程序如下：

```csharp
using System;
using System.Text;
using System.Windows.Forms;
namespace chp04_08_06
{
    public partial class Form1 : Form
    {
        public Form1()
        {
            InitializeComponent();
        }
        private void 打开ToolStripMenuItem_Click(object sender, EventArgs e)
        {
            openFileDialog1.Filter = "txt文件(*.txt)|*.txt";
            if(openFileDialog1.ShowDialog() == DialogResult.OK)
            {
                System.IO.StreamReader sr =
                    new System.IO.StreamReader(openFileDialog1.FileName, Encoding.Default);
                richTextBox1.Text = sr.ReadToEnd();
                sr.Close();
            }
```

```csharp
        }
        private void toolStripButton1_Click(object sender, EventArgs e)
        {
            打开ToolStripMenuItem_Click(sender,e);
        }
        private void 保存ToolStripMenuItem_Click(object sender, EventArgs e)
        {
            saveFileDialog1.Filter = "*.txt|*.txt";
            if (saveFileDialog1.ShowDialog() == DialogResult.OK)
            {
                System.IO.StreamWriter sw =
                System.IO.StreamWriter sw = new System.IO.StreamWriter
                    (saveFileDialog1.FileName,false,System.Text.Encoding.Default);
                sw.Write(richTextBox1.Text);
                sw.Close();
            }
        }
        private void toolStripButton2_Click(object sender, EventArgs e)
        {
            保存ToolStripMenuItem_Click(sender, e);
        }
        private void 字体ToolStripMenuItem_Click(object sender, EventArgs e)
        {
            if (fontDialog1.ShowDialog() == DialogResult.OK)
                richTextBox1.SelectionFont = fontDialog1.Font;
        }
        private void toolStripButton3_Click(object sender, EventArgs e)
        {
            字体ToolStripMenuItem_Click(sender, e);
        }
        private void 颜色ToolStripMenuItem_Click(object sender, EventArgs e)
        {
            if (colorDialog1.ShowDialog() == DialogResult.OK)
                richTextBox1.SelectionColor = colorDialog1.Color;
        }
        private void toolStripButton4_Click(object sender, EventArgs e)
        {
            颜色ToolStripMenuItem_Click(sender, e);
        }
    }
}
```

习题 9

一、选择题

1. A　2. D　3. D　4. D　5. B　6. C　7. D　8. B　9. B　10. C
11. B　12. A

二、填空题

1. 随机文件，二进制文件
2. File，Create，Copy，Delete
3. Directory，CreateDirectory，Delete
4. FileStream，StreamReader，StreamWriter
5. BinaryReader，BinaryWriter
6. Length，Position
7. System.IO，using System.IO；
8. OpenRead
9. EndOfStream，true
10. WriteLine

三、阅读程序题

1. 程序运行结果如下：

```
File 'e:\data\test.txt' has created successful!
Folder name:e:\data\test.txt
size in bytes 0
Lets use this file now.
```

2. 当 E 盘中存在 data 文件夹时的输出结果：

```
Hello,World!
Hello,C#
```

当 E 盘中不存在 data 文件夹时的输出结果：

```
The file can't be written
未能找到路径"e:\data\text1.txt"的一部分.
The file can't be read
未能找到路径"e:\data\text1.txt"的一部分。
```

3. abcdefghijklmnopqrstuvwxyz

四、程序填空题

1. ① new FileInfo(str)   ② file1.CreateText()   ③ file2.Exists   ④ path1，true
2. ① StreamWriter(@"e:\data\t2.txt"，true)；   ② strline = rstream0.ReadLine()
   ③ strline!=null   ④ wstream.Close()

五、编写程序题

1. 参考程序如下：

```csharp
using System;
using System.IO;
namespace chp04_09_06
{
    class opfile
    {
        public string mypath1 = @"e:\data\text01.txt";
        public string mypath2 = @"e:\data\text02.txt";
        public void writefile()
        {
            if (!File.Exists(mypath1))
            {
                StreamWriter wstream1 = File.CreateText(mypath1);
                wstream1.WriteLine("C#程序设计基础");
```

```csharp
            wstream1.Close();
        }
    }
    public void readfile(out string f1)
    {
        StreamReader rstream = File.OpenText(mypath1);
        string s1 = "";
        f1 = "";
        while ((s1 = rstream.ReadLine())!= null)
            f1 = f1 + s1;
        rstream.Close();
    }
    public void copyfile(string path1,string path2)
    {
        if(!File.Exists(mypath2))
        {
            StreamWriter wstream1 = File.CreateText(path2);
            wstream1.Close();
        }
        File.Copy(path1, path2, true);
    }
    public void deletefile(string path2)
    {
        File.Delete(path2);
    }
}
class Program
{
    static void Main(string[] args)
    {
        string s1;
        opfile file1 = new opfile();
        file1.writefile();
        file1.readfile(out s1);
        Console.WriteLine(s1);
        file1.copyfile(file1.mypath1, file1.mypath2);
        file1.deletefile(file1.mypath1);
    }
}
```

2. 创建一个 Windows 程序，在窗体上添加一个 TreeView 控件和一个 ListView 控件。将 TreeView 控件的 Name 属性设置为 treeView；Dock 属性设置为 Left。将 ListView 控件的 Name 属性设置为 listView；View 属性设置为 Detail；Dock 属性设置为 Fill。

分别为窗体的 Load 事件、treeView 的 AfterExpand 事件及 treeView 的 AfterSelect 事件编写事件代码。在 Load 事件中用到了自定义方法（getSubNode 方法），用于获取子目录以创建目录树节点。

程序运行界面如图 4-30 所示。

图 4-30 树形视图和列表视图控件应用程序运行界面

参考程序如下：

```csharp
using System;
using System.IO;
using System.Windows.Forms;
namespace chp04_09_07
{
    public partial class Form1 : Form
    {
        public Form1()
        {
            InitializeComponent();
        }
        private void Form1_Load(object sender, EventArgs e)
        {
            listView.Columns.Add("名称");
            listView.Columns.Add("大小");
            listView.Columns.Add("修改时间");
            string[] MyDrivers = Directory.GetLogicalDrives();
            TreeNode[] croot = new TreeNode[MyDrivers.Length];
            for(int i = 0;i < MyDrivers.Length;i++)
            {
                TreeNode drivesNode = new TreeNode(MyDrivers[i]);
                treeView.Nodes.Add(drivesNode);
                getSubNode(drivesNode, true);
            }
        }
        private void getSubNode(TreeNode pathName,bool isEnd)
```

```csharp
{
    if (!isEnd)   return;
    TreeNode curNode;
    DirectoryInfo[] subDir;
    DirectoryInfo curDir = new DirectoryInfo(pathName.FullPath);
    try
    {
        subDir = curDir.GetDirectories();
        foreach(DirectoryInfo d in subDir)
        {
            curNode = new TreeNode(d.Name);
            pathName.Nodes.Add(curNode);
            getSubNode(curNode, false);
        }
    }
    catch(Exception ex)
    {
        MessageBox.Show(ex.ToString());
    }
}
private void treeView_AfterExpand(object sender, TreeViewEventArgs e)
{
    try
    {
        foreach (TreeNode tn in e.Node.Nodes)
            if (!tn.IsExpanded)
                getSubNode(tn, true);
    }
    catch(Exception ex)
    { MessageBox.Show(ex.ToString()); }
}
private void treeView_AfterSelect(object sender, TreeViewEventArgs e)
{
    listView.Items.Clear();
    DirectoryInfo selDir = new DirectoryInfo(e.Node.FullPath);
    DirectoryInfo[] listDir;
    FileInfo[] listFile;
    try
    {
        ListViewItem mylistvItem;
        ListViewItem.ListViewSubItem mylistVSubItem;
        listDir = selDir.GetDirectories();
        listFile = selDir.GetFiles();
        foreach(DirectoryInfo d in listDir)
        {
            mylistvItem = new ListViewItem();
            mylistvItem.Text = d.Name;
```

```
                    mylistvItem.Tag = d.FullName;
                    mylistVSubItem = new ListViewItem.ListViewSubItem();
                    mylistVSubItem.Text = "";
                    mylistvItem.SubItems.Add(mylistVSubItem);
                    mylistVSubItem = new ListViewItem.ListViewSubItem();
                    mylistVSubItem.Text = d.LastWriteTime.ToString();
                    mylistvItem.SubItems.Add(mylistVSubItem);
                    listView.Items.Add(mylistvItem);
                }
                foreach(FileInfo d in listFile)
                {
                    mylistvItem = new ListViewItem();
                    mylistvItem.Text = d.Name;
                    mylistvItem.Tag = d.FullName;
                    mylistVSubItem = new ListViewItem.ListViewSubItem();
                    mylistVSubItem.Text = "";
                    mylistvItem.SubItems.Add(mylistVSubItem);
                    mylistVSubItem = new ListViewItem.ListViewSubItem();
                    mylistVSubItem.Text = d.LastWriteTime.ToString();
                    mylistvItem.SubItems.Add(mylistVSubItem);
                    listView.Items.Add(mylistvItem);
                }
            }
            catch(Exception ex)
            {
                MessageBox.Show(ex.ToString());
            }
        }
    }
}
```

习题 10

一、选择题

1. D 2. C 3. B 4. A 5. D 6. B 7. D 8. B 9. C 10. A

二、填空题

1. CreateGraphics   2. DataStyle

3. g1.DrawEllipse(Pens.Blue,10,20,100,50);

4. Pen mypen=new Pen(Color.Red,5);

5. Brush

6. SolidBrush mybrush=new SolidBrush(Color.Green);

7. Font，Font myfont = new Font("宋体"，24，FontStyle.Italic);

8. HatchStyle   9. DrawImage

三、阅读程序题

1. 绘制一条贝塞尔曲线，如图 4-31 所示。

2. 单击 button1 按钮后的程序运行界面如图 4-32 所示，显示带阴影的文字。
单击 button2 按钮后的程序运行界面如图 4-33 所示，显示带倾斜效果的文字串。

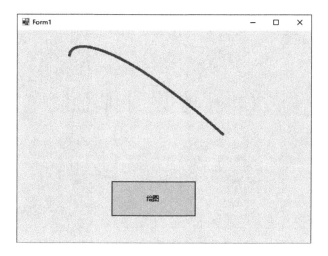

图 4-31　贝塞尔曲线绘制程序运行界面

图 4-32　带阴影文字绘制程序运行界面

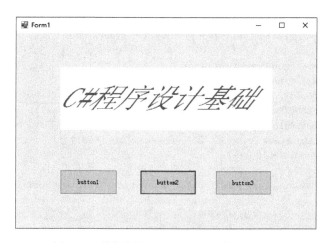

图 4-33　带倾斜效果文字绘制程序运行界面

单击 button3 按钮后的程序运行界面如图 4-34 所示，显示渐变效果的文字串。

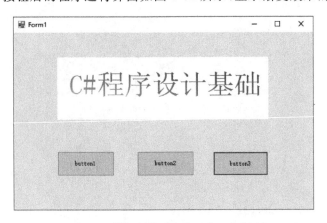

图 4-34　带渐变效果文字程序运行界面

四、程序填空题

1. ① this.CreateGraphics()　　② g.FillRectangle(lBrush，rect)
2. ① Pen(Color.Red)　　　　　② bluePen.EndCap
   ③ mygraphic.DrawLine(bluePen，p1，p2)

五、编写程序题

1. 参考程序如下：

```
using System;
using System.Drawing;
using System.Windows.Forms;
namespace chp04_10_05
{
    public partial class Form1 : Form
    {
        public Form1()
        {
            InitializeComponent();
        }
        private void button1_Click(object sender, EventArgs e)
        {
Pen mypen = new Pen(Color.Black, 1);
Graphics g = this.CreateGraphics();
g.Clear(Color.BurlyWood);
g.DrawImage(Image.FromFile(@"G:\a0.jpg"),10,10,430,480);
g.DrawRectangle(new Pen(Color.Black, 3),
new Rectangle(new Point(20, 20), new Size(410, 460)));
for( int i = 0; i < 10; i++)
{
    g.DrawLine(mypen, new Point(25, 25 + (i * 50)), new Point(425, 25 + (i * 50)));
}
for (int i = 0; i < 9; i++)
{
    g.DrawLine(mypen, new Point(25 + (i * 50),25), new Point(25 + (i * 50),225));
    g.DrawLine(mypen, new Point(25 + (i * 50), 275), new Point(25 + (i * 50), 475));
}
```

```
                g.DrawLine(mypen, new Point(175, 25), new Point(275, 125));
                g.DrawLine(mypen, new Point(275, 25), new Point(175, 125));
                g.DrawLine(mypen, new Point(175, 375), new Point(275, 475));
                g.DrawLine(mypen, new Point(175, 475), new Point(275, 375));
                for(int i = 0;i < 4; i++)
                {
                    g.DrawLine(mypen, new Point(30 + i * 100, 170), new Point(40 + i * 100, 170));
                    g.DrawLine(mypen, new Point(30 + i * 100, 180), new Point(40 + i * 100, 180));
                    g.DrawLine(mypen, new Point(30 + i * 100, 160), new Point(30 + i * 100, 170));
                    g.DrawLine(mypen, new Point(30 + i * 100, 180), new Point(30 + i * 100, 190));
                    g.DrawLine(mypen, new Point(110 + i * 100, 170), new Point(120 + i * 100, 170));
                    g.DrawLine(mypen, new Point(110 + i * 100, 180), new Point(120 + i * 100, 180));
                    g.DrawLine(mypen, new Point(120 + i * 100, 160), new Point(120 + i * 100, 170));
                    g.DrawLine(mypen, new Point(120 + i * 100, 180), new Point(120 + i * 100, 190));
                }
                for (int i = 0; i < 4; i++)
                {
                    g.DrawLine(mypen, new Point(30 + i * 100, 320), new Point(40 + i * 100, 320));
                    g.DrawLine(mypen, new Point(30 + i * 100, 330), new Point(40 + i * 100, 330));
                    g.DrawLine(mypen, new Point(30 + i * 100, 310), new Point(30 + i * 100, 320));
                    g.DrawLine(mypen, new Point(30 + i * 100, 330), new Point(30 + i * 100, 340));
                    g.DrawLine(mypen, new Point(110 + i * 100, 320), new Point(120 + i * 100, 320));
                    g.DrawLine(mypen, new Point(110 + i * 100, 330), new Point(120 + i * 100, 330));
                    g.DrawLine(mypen, new Point(120 + i * 100, 310), new Point(120 + i * 100, 320));
                    g.DrawLine(mypen, new Point(120 + i * 100, 330), new Point(120 + i * 100, 340));
                }
            }
        }
```

2. 程序的设计界面如图 4-35 所示,将 timer1 的 Enabled 属性设置为 true,并编写相应的事件代码和函数。

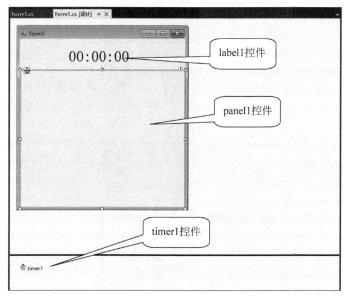

图 4-35 时钟程序设计界面

参考程序如下：

```csharp
using System;
using System.Drawing;
using System.Windows.Forms;
namespace chp04_10_06
{
    public partial class Form1 : Form
    {
        int hh, mm, ss;
        public Form1()
        {
            InitializeComponent();
        }
        private void drawClock(int h, int m, int s)
        {
        Graphics mygraphics = panel1.CreateGraphics();
        mygraphics.Clear(Color.White);
        Font myfont = new Font("黑体", 20, FontStyle.Bold);
        Pen mypen = new Pen(Color.Black, 1);
        SolidBrush mybrush = new SolidBrush(Color.Black);
        mygraphics.DrawString("12", myfont, mybrush, (panel1.Width / 2 - 20), 30);
        mygraphics.DrawString("9", myfont, mybrush, 10,(panel1.Height - 57)/2 + 10);
        mygraphics.DrawString("3", myfont, mybrush,
                panel1.Width - 40, (panel1.Height - 57)/2 + 10);
        mygraphics.DrawString("6", myfont, mybrush,
                (panel1.Width / 2 - 10), (panel1.Height - 70));
        Point centerPoint = new Point(panel1.ClientRectangle.Width/2,
                    panel1.ClientRectangle.Height/2);
        Point secPoint = new Point((int)(centerPoint.X +
                (Math.Sin(s * Math.PI/30) * 120)),(int)(centerPoint.Y -
                (Math.Cos(s * Math.PI/30) * 120)));
        Point minPoint = new Point((int)(centerPoint.X +
                    (Math.Sin(m * Math.PI / 30) * 100)),
                    (int)(centerPoint.Y - (Math.Cos(m * Math.PI / 30) * 100)));
        Point hourPoint = new Point((int)(centerPoint.X +
                    (Math.Sin(h * Math.PI / 6) * 80) - m * Math.PI/360),
                    (int)(centerPoint.Y - (Math.Cos(h * Math.PI / 6) *
                    90) - m * Math.PI/360));
            mypen = new Pen(Color.Red, 1);
            mygraphics.DrawLine(mypen, centerPoint, secPoint);
            mypen = new Pen(Color.Green, 2);
            mygraphics.DrawLine(mypen, centerPoint, minPoint);
            mypen = new Pen(Color.Orange, 4);
            mygraphics.DrawLine(mypen, centerPoint, hourPoint);
        }
        private void timer1_Tick(object sender, EventArgs e)
        {
            label1.Text = DateTime.Now.ToLongTimeString();
            hh = DateTime.Now.Hour;
            mm = DateTime.Now.Minute;
```

```
            ss = DateTime.Now.Second;
            drawClock(hh, mm, ss);
        }
    }
}
```

习题 11

一、选择题
1. A   2. C   3. A   4. B   5. C   6. C   7. D   8. D   9. C   10. D
11. A   12. C   13. A   14. A   15. B

二、填空题
1. Connection，Command，DataAdapter，DataSet
2. System.Data.OleDb
3. ConnectionString，Open，Close
4. OleDbCommand mycmd＝new OleDbCommand("SELECT ＊ FROM student",mycon);
5. OleDbDataReader，OleDbDataReader myreader＝mycmd.ExecuteReader
6. OleDbDataAdapter，DataSet
7. Fill，SelectCommand
8. 数据绑定
9. DataView   10. System.Int16

三、阅读程序题
1. 往 student 表中插入一条记录。
2. 显示 student 表的字段名。

四、程序填空题
1. ① mydt   ② mydt.DefaultView
2. ① new SqlConnection()   ② myconn.Open()   ③ mysql

五、编写程序题
1. 参考程序如下：

```
using System;
using System.Data.SqlClient;
using System.Windows.Forms;
namespace chp04_11_05
{
    public partial class Form1 : Form
    {
        public Form1()
        {
            InitializeComponent();
        }
        private void button1_Click(object sender, EventArgs e)
        {
            DateTime mydate = new DateTime();
            string strcomm = "select * from student where ";
            bool flag1 = false, flag2 = false, flag3 = false;
```

```csharp
SqlConnection mysqlcon = new SqlConnection();
mysqlcon.ConnectionString = "Data Source=DESKTOP-331UI15;
        Initial Catalog=stu;Integrated Security=True";
mysqlcon.Open();
string str1 = textBox4.Text;
if(checkBox1.Checked)
{
    strcomm = strcomm + "id like '" + textBox1.Text.Trim() + "%'";
    flag1 = true;
}
if(checkBox2.Checked)
{
    if (flag1 == true)
        strcomm = strcomm + " and name like '"
            + textBox2.Text.Trim() + "%'";
    else
        strcomm = strcomm + "  name like '"
            + textBox2.Text.Trim() + "%'";
    flag2 = true;
}
if (checkBox3.Checked)
{
    if (flag1 == true || flag2 == true)
    {
        if (int.Parse(textBox3.Text.Trim()) > 0)
            strcomm = strcomm + " and score>="
                + textBox3.Text.Trim();
        if (int.Parse(textBox4.Text.Trim()) > 0)
            strcomm = strcomm + " and score<="
                + textBox4.Text.Trim();
    }
    else
    {
        if (int.Parse(textBox3.Text.Trim()) > 0)
        {
            strcomm = strcomm + " score>="
                + textBox3.Text.Trim();
            flag3 = true;
        };
        if (int.Parse(str1) > 0)
        {
            if (flag3)
                strcomm = strcomm + " and score<="
                    + textBox4.Text.Trim();
            else
                strcomm = strcomm + " score<=" + str1;
        }
    }
}
SqlCommand mysqlcomm = new SqlCommand(strcomm, mysqlcon);
SqlDataReader dr = mysqlcomm.ExecuteReader();
```

```csharp
        listView1.Items.Clear();
        if (dr.HasRows)
        {
            while (dr.Read())
            {
                ListViewItem myitem = new ListViewItem(dr.GetString(0));
                myitem.SubItems.Add(dr.GetString(1));
                myitem.SubItems.Add(dr.GetString(2));
                mydate = dr.GetDateTime(3);
                myitem.SubItems.Add(mydate.ToShortDateString());
                myitem.SubItems.Add(dr[4].ToString());
                listView1.Items.Add(myitem);
            }
        }
        else
            MessageBox.Show("没有满足条件的记录", "提示");
        dr.Close();
        mysqlcon.Close();
    }
    private void checkBox1_CheckedChanged(object sender, EventArgs e)
    {
        if (checkBox1.Checked)
            textBox1.Enabled = true;
        else
            textBox1.Enabled = false;
        textBox1.Text = "";
    }
    private void checkBox2_CheckedChanged(object sender, EventArgs e)
    {
        if (checkBox2.Checked)
            textBox2.Enabled = true;
        else
            textBox2.Enabled = false;
        textBox2.Text = "";
    }
    private void checkBox3_CheckedChanged(object sender, EventArgs e)
    {
        if (checkBox3.Checked)
        {
            textBox3.Enabled = true;
            textBox4.Enabled = true;
            textBox3.Text = "0";
            textBox4.Text = "750";
        }
        else
        {
            textBox3.Enabled = false;
            textBox4.Enabled = false;
            textBox3.Text = "";
            textBox4.Text = "";
        }
```

```
        }
        private void Form1_Load(object sender, EventArgs e)
        {
            this.listView1.Columns.Add("学号", 120, HorizontalAlignment.Left);
            this.listView1.Columns.Add("姓名", 120, HorizontalAlignment.Left);
            this.listView1.Columns.Add("性别", 120, HorizontalAlignment.Left);
            this.listView1.Columns.Add("出生日期", 120,
                              HorizontalAlignment.Left);
            this.listView1.Columns.Add("入学分", 120,
                              HorizontalAlignment.Left);
        }
    }
}
```

2. 参考程序如下：

```
using System;
using System.Data.SqlClient;
using System.Windows.Forms;
namespace chp04_11_06
{
    public partial class Form1 : Form
    {
        string constr = "Data Source=DESKTOP-331UI15;Initial Catalog=
                   stu;Integrated Security=True";
        SqlConnection mycon = new SqlConnection();
        SqlCommand mycom = new SqlCommand();
        public Form1()
        {
            InitializeComponent();
        }
        private void Form1_Load(object sender, EventArgs e)
        {
            mycon.ConnectionString = constr;
            mycom.Connection = mycon;
        }
        private void button1_Click(object sender, EventArgs e)
        {
            mycon.Open();
            mycom.CommandText = "Insert Into Student
                values('00009','蔡洋','男','1997-7-1',700)";
            try
            {
                mycom.ExecuteNonQuery();
                MessageBox.Show("记录已经插入!", "提示");
            }
            catch(Exception ex)
            {
                MessageBox.Show(ex.Message);
            }
            mycon.Close();
```

```csharp
}
private void button2_Click(object sender, EventArgs e)
{
    mycon.Open();
    mycon.CommandText = "Delete from student where sex like '男%'";
    try
    {
        mycom.ExecuteNonQuery();
        MessageBox.Show("记录已经删除!", "提示");
    }
    catch (Exception ex)
    {
        MessageBox.Show(ex.Message);
    }
    mycon.Close();
}
private void button3_Click(object sender, EventArgs e)
{
    mycon.Open();
    mycom.CommandText = "Update student set name = '李闻' 
            where id = '00004'";
    if (mycom.ExecuteNonQuery() > 0)
        MessageBox.Show("记录已更新!", "提示");
    else
        MessageBox.Show("没有满足条件的记录!", "提示");
    mycon.Close();
}
private void button4_Click(object sender, EventArgs e)
{
    mycon.Open();
    mycom.CommandText = "Select min(score) from student";
    int min = Convert.ToInt32(mycom.ExecuteScalar());
    mycom.CommandText = "Select max(score) from student";
    int max = Convert.ToInt32(mycom.ExecuteScalar());
    mycom.CommandText = "Select avg(score) from student";
    int avg = Convert.ToInt32(mycom.ExecuteScalar());
    MessageBox.Show("最低分: " + min.ToString() + "\n最高分: " +
        max.ToString() + "\n平均分: " + avg.ToString(), "统计结果");
}
}
}
```

第5章 模拟试题

本章包括3套C#程序设计的模拟试题和参考答案,涵盖了本课程的主要知识点,可以帮助读者了解和检验自己的学习情况。

模拟试题 1

一、选择题(每题2分,共30分)

1. 在C#中,装箱是把值类型转换到( )类型。
   A. 数组   B. 引用   C. char   D. string

2. 静态构造函数只能对( )数据成员进行初始化。
   A. 静态   B. 动态   C. 实例   D. 静态和实例

3. 下面关于C#的常量定义,正确的是( )。
   A. Const double Pi3.14;   B. Const double n=2.7
   C. define double Pi 3.14   D. define double n=2.7

4. 在Windows应用程序中,如果复选框控件的Checked属性值设置为true,表示( )。
   A. 该复选框被选中   B. 该复选框不被选中
   C. 不显示该复选框的文本信息   D. 显示该复选框的文本信息

5. 面向对象语言的基本特征不包括( )。
   A. 封装   B. 委托   C. 继承   D. 多态

6. 在C#中,从属性的读写操作特性进行分类,可以划分为以下三种,除了( )。
   A. 只读   B. 只写   C. 读写   D. 不可读不可写

7. 以下叙述中正确的是( )。
   A. 接口中可以有虚方法   B. 一个类可以实现多个接口
   C. 接口可以被实例化   D. 接口中可以包含已经实现的方法

8. 关于如下程序结构的描述中,( )是正确的。
   ```
   for ( ; ; )
       s=s+1;
   ```
   A. 不执行语句s=s+1;   B. 一直执行语句s=s+1;即死循环
   C. 执行语句s=s+1一次   D. 程序不符合语法要求

9. 下列( )对象是ADO.NET在非连接模式下处理数据内容的主要对象。
   A. Command   B. Connection   C. DataAdapter   D. DataSet

10. 下列关于数组访问的描述中,(　　)是不正确的。

　　A. 数组元素索引是从 0 开始的

　　B. 对数组元素的所有访问都要进行边界检查

　　C. 如果使用的索引小于 0,或大于数组的大小,编译器将抛出一个 IndexOutOfRangeException 异常

　　D. 数组元素的访问是从 1 开始,到 Length 结束

11. 下列结构图对应于三种程序结构的(　　)(A 是程序段,P 是条件)。

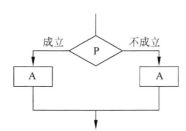

　　A. while 循环结构　　　　　　　B. do-while 循环结构

　　C. if-else 选择结构　　　　　　D. switch-case 选择结构

12. 在 C♯ 的类结构中,class 前面的关键字是表示访问级别,下面(　　)关键字表示该类只能被这个类的成员或派生类成员访问。

　　A. public　　　　B. private　　　　C. internal　　　　D. protected

13. 在 C♯ 中,下列代码运行后,变量 Max 的值是(　　)。

```
Int a = 5, b = 10, c = 15, Max = 0;
Max = a > b?a:b;
Max = c < Max?c:Max;
```

　　A. 0　　　　　　B. 5　　　　　　C. 10　　　　　　D. 15

14. C♯ 中 TestClass 为自定义类,其中有以下属性定义

```
public void Property{ … }
```

使用以下语句创建了该类的对象,并使变量 obj 引用该对象：

```
TestClassobj = new TestClass();
```

那么,可通过(　　)方式访问类 TestClass 的 Property 属性。

　　A. Obj,Property;　　　　　　　B. MyClass.Property;

　　C. obj :: Property;　　　　　　D. obj.Property();

15. 下列关于 C♯ 面向对象应用的描述中,(　　)是正确的。

　　A. 派生类是基类的扩展,派生类可以添加新的成员,也可去掉已经继承的成员

　　B. abstract 方法的声明必须同时实现

　　C. 声明为 sealed 的类不能被继承

　　D. 接口像类一样,可以定义并实现方法

二、填空题(每空 2 分,共 20 分)

1. 如果一个属性里既有 set 访问器又有 get 访问器,那么该属性为_____属性。

2. 要关闭已打开的数据库连接，应使用连接对象的_____方法。
3. 在 do-while 循环结构中，循环体至少要执行_____次。
4. C#支持的循环有 for、while、do-while 和_____循环。
5. 下列程序段执行后，a[3]的值为_____。

int []a = {1,2,3,4,5};a[3] = a[a[3]];

6. 要使 Lable 控件显示给定的文字"您好"，应设置它的_____属性值。
7. 设 x 为 int 型变量，请写出描述"x 是奇数"的 C#语言表达式_____。
8. 在 C#类中，_____是控制台类，利用它可以方便地进行控制台的输入输出。
9. 填充数据集应调用数据适配器的_____方法。
10. 表达式"7==7&&10>4+5"的值为_____。

三、判断题（每题 1 分，共 10 分）

1. 在使用变量之前必须先声明类型。   （   ）
2. 由 static 修饰的成员必须由类来访问而不能通过对象访问。   （   ）
3. 抽象类中所有的方法必须被声明为 abstract。   （   ）
4. 命令对象的 ExecuteScalar()方法是在获取单值的时候使用。   （   ）
5. 下列语句是否正确：   （   ）

for(int i = 0,i<10,i++)Console.WriteLine(i);

6. 属性必须同时定义 get 块和 set 块。   （   ）
7. 类是对象的抽象，对象是类的实例。   （   ）
8. 委托是将方法作为参数传递给另一方法的一种数据类型。事件与委托没有关系。
                                                               （   ）
9. Main 方法的返回值类型必须是 void 的类型。   （   ）
10. C#程序中使用 0 表示逻辑非。   （   ）

四、读程序，写结果（每题 6 分，共 30 分）

1.

```csharp
using System;
namespace chp05_01_01
{
    class Program
    {
        static void Main(string[] args)
        {
            Tiger t = new Tiger();
        }
    }
    class Animal
    {
        public Animal()
        {
            Console.Write("基类");
        }
```

```
        }
        class Tiger : Animal
        {
            public Tiger()
            {
                Console.Write("派生类");
            }
        }
    }
```

2.
```
using System;
namespace chp05_01_02
{
    class Program
    {
        static void Main(string[] args)
        {
            MyClass m = new MyClass();
            int[] s = { 2, 6, 4, 7, 3, 89, 5 };
            int num1, num2;
            m.process(s, out num1, out num2);
            Console.WriteLine("num1 = {0},num2 = {1}", num1, num2);
            Console.ReadLine();
        }
    }
    classMyClass
    {
        public void process(int[] a, out int n1, out int n2)
        {
            n1 = n2 = a[0];
            for (int i = 1; i < a.Length; i++)
            {
                if (a[i] > n1)
                    n1 = a[i];
                if (a[i] < n2)
                    n2 = a[i];
            }
        }
    }
}
```

3.
```
using System;
namespace chp05_01_03
{
    class Program
    {
        static void Main(string[] args)
        {
```

```csharp
            Point p1 = new Point();
            Point p2 = new Point(3, 4);
            Console.WriteLine("p1.x = {0},p1.y = {1}", p1.x, p1.y);
            Console.WriteLine("p2.x = {0:f},p2.y = {1}", p2.x, p2.y);
        }
    }
    class Point
    {
        public double x = 0, y = 0;
        public Point()
        {
            x = 1; y = 1;
        }
        public Point(double a, double b)
        {
            x = a;
            y = b;
        }
    }
}
```

4.
```csharp
using System;
namespace chp05_01_04
{
    class Program
    {
        static void Main(string[] args)
        {
            int i, j;
            int [] a = new int[5] { 1,0,0,0,0};
            for (j = 0; j < 5; j++)
                for (i = 0; i < j; ++i)
                    a[j] = a[j] + a[i];
            Console.WriteLine("a[4] = {0}", a[4]);
        }
    }
}
```

5.
```csharp
using System;
namespace chp05_01_05
{
    class Program
    {
        static void Main(string[] args)
        {
            int i = 0, sum = 0;
            do
            {
```

```
                sum++;
        } while (i > 0);
        Console.WriteLine("sum = {0}", sum);
    }
}
```

五、编程题(10分)

如果一个数恰好等于它的真因子(除了自身的约数)之和,则称该数为"完全数"。例如对于自然数6,它的真因子是1、2、3,而1+2+3=6,因此6是完全数。编程求1000以内的完全数。

模拟试题2

一、选择题(每题2分,共30分)

1. 下列标识符命名正确的是(　　)。
   A. X.25　　　　　　B. 4foots　　　　　C. val(7)　　　　　D. _Years
2. 以下说法正确的是(　　)。
   A. 构造函数名不必和类名相同
   B. 一个类可以声明多个构造函数
   C. 构造函数可以有返回值
   D. 编译器可以提供一个默认的带一个参数的构造函数
3. 以下说法不正确的是(　　)。
   A. 一个类可以实现多个接口
   B. 一个派生类可以继承多个基类
   C. 在C#中实现多态,派生类中重写基类的虚函数必须在前面加override
   D. 子类能添加新方法
4. C#程序中的类myClass中的成员变量var1如下:

```
public class myClass{
privateint var1;
}
```

则可以访问var1的有(　　)。
   A. 在myClass类中　　　　　　　　B. myClass的继承类
   C. myClass的父类　　　　　　　　D. 和myClass在同一命名空间下的类
5. 以下正确的描述是(　　)。
   A. 函数的定义可以嵌套,函数的调用不可以嵌套
   B. 函数的定义不可以嵌套,函数的调用可以嵌套
   C. 函数的定义和函数的调用均可以嵌套
   D. 函数的定义和函数的调用均不可以嵌套
6. 下面关于抽象类的说法正确的是(　　)。
   A. 抽象类可以被实例化

B. 含有抽象方法的类一定是抽象类

C. 抽象类可以是静态类和密封类

D. 抽象类中的抽象方法可以在其他类中重写

7. 异常捕获发生在（　　）块中？

  A. try   B. catch   C. finally   D. throw

8. 变量定义如下：

```
int a = 19,b = 6,c = 7,d = 5;
bool s,e = false;
```

则表达式(s＝a＜c)＆＆(e＝b-d＞0)运算后，e的值是（　　）。

  A. 0   B. 1   C. true   D. false

9. 在类的定义中，类的（　　）描述了该类的对象的行为特征。

  A. 类名   B. 方法   C. 所属的命名空间 D. 私有域

10. 在 C# 中，下列代码的运行结果是（　　）。

```
int[] num = new int[] { 1, 3, 2, 0, 0 };
Array.Reverse(num);
foreach (int i in num)
Console.Write(i);
```

  A. 00123   B. 12300   C. 00231   D. 13200

11. 在 C# 语言的循环结构中，首先执行一次，然后再判断条件的循环结构是（　　）。

  A. while 循环   B. do-while 循环

  C. for 循环   D. foreach 循环

12. 以下关于 C# 代码的说法正确的是（　　）。

```
for (int i = 1; i <= 3; i++)
{
    switch (i)
    {
        case 1:Console.Write(i.ToString());
        case 2:
        Console.Write((i * 2).ToString());
        case 3:
        Console.Write((i * 3).ToString());
    }
}
```

  A. 有编译错误，提示 case 标签不能贯穿到另一个标签，不能运行

  B. 输出 149

  C. 输出 123246369

  D. 正确运行，但没有输出

13. 在 C# 中，下列代码的运行结果是（　　）。

```
using System;
class Test{
    static Void Main(string [] args){
```

```
        int a = 5,b = 9;
        int c = a > b?a++: -- b;
        Console.WriteLine(c);
    }
}
```

    A. 5   B. 9   C. 8   D. 6

14. 委托声明的关键字是（　　）。

    A. delegate   B. delete   C. public   D. interface

15. 当整数 a 赋值给一个 object 对象时，整数 a 将会被（　　）。

    A. 拆箱   B. 丢失   C. 装箱   D. 出错

二、填空题（每空 2 分，共 20 分）

1. 面向对象语言都应至少具有的三个特性是：封装、_____和多态。
2. 在循环结构中，continue 语句的作用是_____。
3. 在 C♯ 程序中，程序的执行总是从_____方法开始的。
4. 在 C♯ 中，进行注释有两种方法：使用_____和使用"/＊　＊/"符号对，其中前者只能进行单行注释。
5. 定义方法时使用的参数是形参，调用方法时使用的参数是_____。
6. 下列程序段执行后，a[3]的值为_____。

`int []a = {1,2,3,4,5};a[3] = a[a[3] - 3];`

7. 当在程序中执行到_____语句时，将结束本层循环类语句或 switch 语句的执行。
8. 在 C♯ 中创建一个对象时，会先执行该对象的_____中的语句。
9. 在 ADO.NET 中，建立到数据库的连接的对象为_____对象。
10. 按钮控件的常用事件是_____事件。

三、判断题（每题 1 分，共 10 分）

1. C♯ 中所有的类型实质上都是从 object 类派生而来的。（　　）
2. 在同一行上可以书写多条语句，每条语句间用分号分隔。（　　）
3. 类和对象的区别就是，类是对象的实例，而对象则是类的抽象。（　　）
4. 以下的代码执行后，string 类型变量 a 的值是 changed。（　　）

```
static private void b(string c)
{
    c = "changed";
}
static void Main()
{
    string a = "original value";
    b(a);
}
```

5. for 循环只能用于循环次数已经确定的情况。（　　）
6. 子类可以从父类中继承其所有的成员。（　　）
7. 当程序中只需要进行两个选择，一般采用 if-else 语句。（　　）

8. 在C#中,声明一个变量intabc,不给abc赋值直接使用abc,它将获得默认值0。
( )

9. 在一个程序内,不可以包含2个及以上的Main方法。
( )

10. 在C#中,接口可以被多重继承而类不能。
( )

四、读程序,写结果(每题6分,共30分)

1.

```
using System;
namespace chp05_02_01
{
    class Program
    {
        static void Main(string[] args)
        {
            int i, j, m, n;
            int[,] a = new int[3, 4] { { 32, 11, 3, 12 }, { 15, 12, 6, 43 },
                                       { 29, 19, 21, 11 } };
            m = n = 0;
            for(i = 0;i < 3;i++)
                for(j = 0;j < 4;j++)
                    if (a[m, n] > a[i, j])
                    {
                        m = i;n = j;
                    }
            Console.WriteLine("a[{0},{1}] = {2}", m,n,a[m, n]);
        }
    }
}
```

2.

```
using System;
namespace chp05_02_02
{
    class Program
    {
        staticint fun(int h)
        {
            int[] a = new int[3] { 1, 2, 3 };
            int k;
            for (k = 0; k < 3; k++)
                a[k] += a[k] - h;
            for (k = 0; k < 3; k++)
                Console.Write("{0}    ",a[k]);
            return a[h];
        }
        static void Main(string[] args)
        {
            int t = 0;
            fun(fun(t));
```

            }
        }
    }

3.

```
using System;
namespace chp05_02_03
{
    class Program
    {
        static void Main(string[] args)
        {
            s s1 = new s();
            s t1 = new s();
        }
    }
    public class s
    {
        public s()
        {
            Console.Write("构造函数!");
        }
        static s()
        {
            Console.Write("静态构造函数!");
        }
    }
}
```

4.

```
using System;
namespace chp05_02_04
{
    class Program
    {
        static void Main(string[] args)
        {
            Derive d = new Derive();
            Base b = d;
            b.G();
            d.G();
        }
    }
    class Base
    {
        public virtual void G()
        {
            Console.Write("Base.G!");
        }
```

```csharp
    class Derive : Base
    {
        public override void G()
        {
            Console.Write("Derive.G!");
        }
    }
}
```

5.
```csharp
using System;
namespace chp05_02_05
{
    class Program
    {
        static void Main(string[] args)
        {
            int j;
            for (int i = 0; i <= 9; i++)
            {
                j = i * 10 + 6;
                if (i % 10 != 6 && j % 3 != 0)
                    continue;
                Console.Write("\t{0}", j);
            }
        }
    }
}
```

五、编程题(10 分)

已知数列 1,3,3,3,5,5,5,5,5,7,7,7,7,7,7,7……编程求：第 40 项的值；值为 17 的第一个数是数列中的第几项。

模拟试题 3

一、选择题(每题 2 分,共 30 分)

1. 从值类型转换到引用类型称为(　　)。

　　A. 继承　　　　B. 拆箱　　　　C. 装箱　　　　D. 转换

2. 对于运算符 & 与 &&,描述正确的是(　　)。

　　A. 二者含义一样,可以通用

　　B. & 是位运算符,表示按位与运算,&& 是逻辑与运算符

　　C. 都可以进行关系运算

　　D. 都可以进行逻辑运算

3. 面向对象编程中,"继承"的概念是指(　　)。

　　A. 对象之间通过消息进行交互

　　B. 派生自同一个基类的不同类的对象具有一些共同特征

C. 对象的内部细节被隐藏

D. 派生类对象可以不受限制地访问所有的基类对象

4. C#中 MyClass 为一自定义类，其中有以下方法定义

public void Hello(){ -- }

使用以下语句创建了该类的对象，并使变量 obj 引用该对象：

MyClassobj = new MyClass();

那么，可访问类 MyClass 的 Hello 方法的语句是(　　)。

A. obj.Hello();　　　　　　　　B. obj::Hello();

C. MyClass.Hello();　　　　　　D. MyClass::Hello();

5. 分析下列代码段，运行结果是(　　)。

```
static void Main(string[] args)
{
    string[] words = new string[] {"a","b","c"};
    foreach (string word in words)
    {
        word = "abc";
        Console.Write (word);
    }
}
```

A. abc

B. abcabcabc

C. bc ac ab

D. 不能正确编译

6. 下面属于合法变量名的是(　　)。

A. P_qr　　　B. 123mnp　　　C. char　　　D. x-y

7. 下面代码的运行结果是(　　)。

```
static void Main(string[ ] args)
{
    int num1 = 34;
    int num2 = 55;
    Increase(ref num1, num2);
    Console.WriteLine("{0}和{1}", num1, num2);
    Console.ReadLine();
}
private static void Increase (ref int num1, int num2)
{
    num1++;
    num2++;
}
```

A. 35 和 56　　B. 34 和 55　　C. 34 和 56　　D. 35 和 55

8. 表达式 12/4－2＋5＊8/4％5/2 的值为(　　)。

A. 1　　　　B. 3　　　　C. 4　　　　D. 10

9. 以下叙述正确的是（   ）。
   A. 一个类可以实现多个接口   B. 接口中可以有虚方法
   C. 接口中可以包含已实现的方法   D. 接口可以被实例化

10. 在 C# 的控件中，Panel、GroupBox 和 TabControl 等分组控件有时候也被称为（   ）。
    A. 组合控件   B. 容器控件   C. 基类控件   D. 排列控件

11. .NET 框架是.NET 战略的基础，是一种新的便捷的开发平台，它具有两个主要的组件，分别是（   ）和类库。
    A. 公共语言运行库   B. Web 服务
    C. 命名空间   D. Main()函数

12. 下列能正确创建数组的是（   ）。
    A. int[,] array=int[4,5];
    B. int size=int.Parse(Console.ReadLine());
       int[] pins=new int [size];
    C. string[] str=new string[];
    D. intpins[] = new int[2];

13. 下面关于抽象类的说法正确的是（   ）。
    A. 抽象类只能做子类   B. 抽象类可以被实例化
    C. 抽象类不能被实例化   D. 一个抽象类只能有一个子类

14. 在 C# 语言中，下列异常处理结构中有错误的是（   ）。
    A. catch{ }   finally{ }   B. try{ } finally{ }
    C. try{} catch{ } finally{ }   D. try{ } catch{ }

15. 关于 C# 中的 switch-case 语句，以下说法正确的是（   ）。
    A. switch 判断的表达式可以是整型或者字符型，但不能是字符串型
    B. 在该语句中最多不能超过 5 个 case 子句
    C. 在该语句中只能有一个 default 子句
    D. 在该句中只能有一个 break 语句

二、填空题（每空 2 分，共 20 分）

1. C# 通过_____和拆箱机制，可以实现值类型和引用类型之间的转换。
2. 在 DataSet 对象中，可通过_____集合遍历 DataSet 对象中所有的数据表对象。
3. 在类的成员声明时，若使用了_____修饰符，则该成员只能在该类或其派生类中使用。
4. _____是具有相同或相似性质的对象的抽象。
5. 表达式"8%3－2==1"的值是_____。
6. 常量被声明为字段，声明时在字段的类型前面使用_____关键字。
7. 在 C# 中实参与形参有 4 种传递方式，它们分别是_____、引用参数、输出参数和参数数组。
8. StreamWriter 的_____方法，可以向文本文件写入一行带回车和换行的文本。
9. 在 C# 程序中，显示一个信息为"This is a test!"，标题为 Hello 的消息框，语句

是_____。

10. 定义一个由 3 个元素组成的交错数组,其语句是_____。

三、判断题(每题 1 分,共 10 分)

1. 可以不使用 new 关键字来对数组进行初始化。   (   )
2. 布尔型变量可以赋值为 0 或 1。   (   )
3. 结构和类均为引用类型。   (   )
4. foreach 语句既可以用来遍历数组中的元素,又可以改变数据元素的值。   (   )
5. 在有继承关系的类中,当创建派生类的对象时,先调用派生类的构造函数,再调用基类构造函数。   (   )
6. 构造函数和析构函数均可以被显式调用。   (   )
7. .NET 包含两个部分,即公共语言运行库和框架类库。   (   )
8. 枚举型是值类型,它是一组称为枚举数列表的命名常量组成的独特类型。   (   )
9. 基类中被说明为 protected 和 private 的成员只能被其派生类的成员函数访问,不能被其他的函数访问。   (   )
10. 如果使用的整数索引小于 0,或者大于数组的大小,编译器将抛出一个 IndexOutOfRangeException 异常。   (   )

四、读程序,写结果(每题 6 分,共 30 分)

1. 运行下列程序,从键盘分别输入 12 和 8,写出运行结果。

```
using System;
namespace chp05_03_01
{
    class Program
    {
        static void Main(string[] args)
        {
            int m, n, i, j, max = 0;
            Console.WriteLine("请输入 m,n 的值");
            m = Convert.ToInt32(Console.ReadLine());
            n = int.Parse(Console.ReadLine());
            if (m < n)
                i = m;
            else
                i = n;
            for (j = i; j > 0; j--)
                if (m % j == 0 && n % j == 0)
                {
                    max = j;
                    break;
                }
            Console.WriteLine("max = {0}", max);
        }
    }
}
```

2.
```csharp
using System;
namespace chp05_03_02
{
    class Program
    {
        static void Main(string[] args)
        {
            int Sum = 0;
            for (int i = 1; i <= 8; i++)
            {
                if (i % 2 == 0)
                    Sum += i;
            }
            Console.WriteLine(Sum);
        }
    }
}
```

3.
```csharp
using System;
namespace chp05_03_03
{
    class Example
    {
        publicint a;
        static public int b;
        public void meth1()
        {
            a = 15;
            b = 25;
        }
        public static void meth2()
        {
            b = 35;
        }
    }
    class Program
    {
        static void Main(string[] args)
        {
            Example e1 = new Example();
            e1.meth1();
            Example.meth2();                    //调用 meth2()
            Console.WriteLine("a = {0},b = {1}", e1.a, Example.b);
        }
    }
}
```

4.
```
using System;
namespace chp05_03_04
{
    class Program
    {
        static void Main(string[] args)
        {
            try
            {
                int a = 10;
                int b = 0;
                int c = a / b;
                Console.WriteLine(c);
            }
            catch
            {
                Console.WriteLine("出现错误");
            }
            finally
            {
                Console.WriteLine("运行结束");
            }
        }
    }
}
```

5.
```
using System;
namespace chp05_03_05
{
    class Program
    {
        static void Main(string[] args)
        {
            int N = 6,t,i,j;
            int[] a = { 1, 2, 3, 4, 5, 6 };
            i = 0;
            j = N - 1;
            for(;i < j; i++, j--)
            {
                t = a[i]; a[i] = a[j];a[j] = t;
            }
            for (i = 0; i < N; i++)
                Console.Write("{0}\t", a[i]);
        }
    }
}
```

五、编程题

设 n 是一任意自然数，若将 n 的各位数字反向排列所得自然数 n1 与 n 相等，则称 n 为回文数。例如，若 n=1234321，则称 n 为回文数；但若 n=1234567，则 n 不是回文数。从键盘上输入一个正整数，判别它是否为回文数。

参 考 答 案

模拟试题 1

一、选择题

1. B   2. A   3. B   4. A   5. B   6. D   7. B   8. B   9. D   10. D
11. C   12. D   13. C   14. A   15. C

二、填空题

1. 读写   2. Close   3. 1   4. foreach   5. 5
6. Text   7. x%2==1   8. Console   9. Fill   10. true

三、判断题

1. √   2. √   3. ×   4. √   5. ×   6. ×   7. √   8. ×   9. ×   10. ×

四、读程序，写结果

1. 基类派生类

2. num1=89,num2=2

3. p1.x=3,p1.y=2
   p2.x=5.00,p2.y=8

4. a[4]=8

5. sum = 1

五、编程题

参考程序如下：

```csharp
using System;
namespace chp05_01_06
{
    class Program
    {
        static void Main(string[] args)
        {
            int i, j, s;
            for (i = 1; i <= 1000; i++)
            {
                s = 0;
                for (j = 1; j <= i / 2; j++)
                    if(i % j == 0)
                        s = s + j;
                if (i == s)
                    Console.WriteLine(i);
            }
```

            }
        }
}

模拟试题 2

一、选择题

1. D　 2. B　 3. B　 4. A　 5. B　 6. B　 7. B　 8. D　 9. B　 10. C
11. B　 12. A　 13. C　 14. A　 15. C

二、填空题

1. 继承　 2. 结束本次循环，继续下一次循环　 3. Main　 4. //　 5. 实参
6. 2　 7. break　 8. 构造函数　 9. Connection　 10. Click 或单击

三、判断题

1. √　 2. √　 3. ×　 4. ×　 5. ×　 6. ×　 7. √　 8. ×　 9. √　 10. √

四、读程序，写结果

1. a[0,2]=3
2. 2　4　6　0　2　4
3. 静态构造函数！构造函数！构造函数！
4. Derive.G！Derive.G！
5. 6　　36　　66　　96

五、编程题

参考程序一：

```
using System;
namespace chp05_02_06
{
    class Program
    {
        static void Main(string[] args)
        {
            int[] a = new int[100];
            int i = 0, j = 1, s = 0;
            while (i <= 99)
            {
                s = s + 1;
                for (j = 1; j <= 2 * s - 1; j++)
                {
                    a[i] = 2 * s - 1;
                    i = i + 1;
                    if (i > 100)
                        break;
                }
            }
            Console.WriteLine("the 40th of the number is {0}", a[39]);
            for (i = 0; i < 100; i++)
                if (a[i] == 17)
                {
```

```
                Console.WriteLine("the first value of 17's position is {0}",
                                   i + 1);
                break;
            }
        }
    }
}
```

参考程序二：

```
using System;
namespace chp05_02_07
{
    class Program
    {
        static void Main(string[] args)
        {
            int i = 0, j, s = 0;
            bool flag = true;
            while (flag)
            {
                s = s + 1;
                for (j = 1; j <= 2 * s - 1; j++)
                {
                    i++;
                    if (i == 40)
                        Console.WriteLine(2 * s - 1);
                    if (2 * s - 1 == 17 && flag)
                    {
                        Console.WriteLine(i);
                        flag = false;
                    }
                }
            }
        }
    }
}
```

模拟试题 3

一、选择题

1. C　2. B　3. B　4. A　5. D　6. A　7. D　8. A　9. A　10. B
11. A　12. B　13. C　14. A　15. C

二、填空题

1. 装箱　2. Tables　3. protected　4. 类　5. false　6. const
7. 值参数　8. WriteLine　9. MessageBox.Show("this is a test!","Hello")；
10. int [][] a=new int[3][]；

三、判断题

1. √　2. ×　3. ×　4. ×　5. ×　6. ×　7. √　8. √　9. ×　10. √

四、读程序，写结果

1. max＝4
2. 20
3. a＝15,b＝35
4. 出现错误
   运行结束
5. 6　　5　　　4　　　3　　　2　　　1

五、编程题

参考程序如下：

```
using System;
namespace chp05_03_06
{
    class Program
    {
        static void Main(string[] args)
        {
            long num, t, s = 0;
            num = long.Parse(Console.ReadLine());
            t = num;
            while (t != 0)
            {
                s = s * 10 + t % 10;
                t = t / 10;
            }
            if (s == num)
                Console.WriteLine("{0}是回文数", num);
            else
                Console.WriteLine("{0}不是回文数", num);
        }
    }
}
```

参 考 文 献

[1] 李春葆,曾平,喻丹丹. C♯程序设计教程[M]. 3版. 北京:清华大学出版社,2015.
[2] 高国江,高凯. C♯程序开发习题解析及实验教程[M]. 北京:清华大学出版社,2014.
[3] 赵敏. C♯程序设计基础——教程、实验、习题[M]. 北京:电子工业出版社,2011.
[4] 江红,余青松. C♯程序设计教程[M]. 2版. 北京:清华大学出版社,2014.
[5] 姜桂洪,等. Visual C♯.NET程序设计实践与题解[M]. 北京:清华大学出版社,2011.
[6] 郑阿奇. C♯实用教程[M]. 2版. 北京:电子工业出版社,2013.
[7] 张世明. C♯程序设计基础[M]. 北京:电子工业出版社,2016.
[8] BenjaminPerkins,Jacob Vibe Hammer,JonD Reid. C♯入门经典[M]. 7版. 齐立波,黄俊伟,译. 北京:清华大学出版社,2016.
[9] 刘军,刘瑞新. C♯程序设计教程上机指导与习题解答[M]. 北京:机械工业出版社,2012.